"In response to the traumas of climate catastrophe, Lawson's *Ecological Ethics and the Philosophy of Simone Weil* shows us that suffering and beauty can be integrated at the heart of environmental consciousness. Like Keller's *Face of the Deep* and Leopold's *Sand County Almanac*, this is a rare treasure that unites profound intellectual insight and ethical urgency."

Daniel O'Dea Bradley, *Professor of Philosophy, Gonzaga University, USA*

Ecological Ethics and the Philosophy of Simone Weil

This book places the philosophy of Simone Weil into conversation with contemporary environmental concerns in the Anthropocene.

The book offers a systematic interpretation of Simone Weil, making her ethical philosophy more accessible to non-Weil scholars. Weil's work has been influential in many fields, including politically and theologically based critiques of social inequalities and suffering, but rarely linked to ecology. Kathryn Lawson argues that Weil's work can be understood as offering a coherent approach with potentially widespread appeal applicable to our ethical relations to much more than just other human beings. She suggests that the process of "decreation" in Weil is an expansion of the self which might also come to include the surrounding earth and a vast assemblage of others. This allows readers to consider what it means to be human in this time and place, and to contemplate our ethical responsibilities both to other humans and also to the more-than-human world. Ultimately, the book uses Weil's thought to decenter the human being by cultivating human actions towards an ecological ethics.

This book will be useful for Simone Weil scholars and academics, as well as students and researchers interested in environmental ethics in departments of comparative literature, theory and criticism, philosophy, and environmental studies.

Kathryn Lawson is a lecturer of philosophy at Carleton University in Ottawa, Canada. She is co-editor of *Hannah Arendt and Simone Weil: Unprecedented Conversations* (2024) and *Breached Horizons: The Philosophy of Jean-Luc Marion* (2017) and author of a number of peer-reviewed journal articles and chapters.

Routledge Environmental Ethics
Series Editor: Benjamin Hale, University of Colorado, Boulder

The Routledge Environmental Ethics series aims to gather novel work on questions that fall at the intersection of the normative and the practical, with an eye toward conceptual issues that bear on environmental policy and environmental science. Recognizing the growing need for input from academic philosophers and political theorists in the broader environmental discourse, but also acknowledging that moral responsibilities for environmental alteration cannot be understood without rooting themselves in the practical and descriptive details, this series aims to unify contributions from within the environmental literature.

Books in this series can cover topics in a range of environmental contexts, including individual responsibility for climate change, conceptual matters affecting climate policy, the moral underpinnings of endangered species protection, complications facing wildlife management, the nature of extinction, the ethics of reintroduction and assisted migration, reparative responsibilities to restore, among many others.

For more information on the series, please visit https://www.routledge.com/Routledge-Environmental-Ethics/book-series/ENVE

Climate Justice Beyond the State
Lachlan Umbers and Jeremy Moss

The Concept of Milieu in Environmental Ethics
Individual Responsibility within an Interconnected World
Laÿna Droz

Critical Realism and the Objective Value of Sustainability
Philosophical and Ethical Approaches
Gabriela-Lucia Sabau

Ecological Ethics and the Philosophy of Simone Weil
Decreation for the Anthropocene
Kathryn Lawson

Ecological Ethics and the Philosophy of Simone Weil

Decreation for the Anthropocene

Kathryn Lawson

Routledge
Taylor & Francis Group
LONDON AND NEW YORK

earthscan
from Routledge

First published 2024
by Routledge
4 Park Square, Milton Park, Abingdon, Oxon, OX14 4RN

and by Routledge
605 Third Avenue, New York, NY 10158

Routledge is an imprint of the Taylor & Francis Group, an informa business

British Library Cataloguing-in-Publication Data
A catalogue record for this book is available from the British Library

ISBN: 978-1-032-58329-7 (hbk)
ISBN: 978-1-032-58326-6 (pbk)
ISBN: 978-1-003-44962-1 (ebk)

DOI: 10.4324/9781003449621

Typeset in Times New Roman
by Newgen Publishing UK

For five extraordinary women:

Erin Lawson, Elaine Ferrier, Alexis Pooley, Charlotte Usselman, and Simone Weil.

For Simone Weil, the catalyst of this work. For my constant comrades in the work of life: Elaine, Alexis, and Charlotte. And for my Mom, Erin, who remains my biggest supporter after all these years.

Contents

Figures

Acknowledgments

I have been overwhelmed by the kindness and support of Simone Weil scholars across the world including an incredible group of academics from the American Weil Society: Ronald Collins, who invited me to work with him on his magnificent journal, *Attention: The Life and Legacy of Simone Weil*, Eric O. Springsted, who generously answered my questions and helped challenge my ideas about Weil, and especially E. Jane Doering, who continually made time for my work and ideas and offered insightful feedback. Scott Ritner and Ben Davis have also offered much-appreciated support and insight. Simone Kotva has been an incredible source of assistance and knowledge, particularly during my time working with her at Cambridge, and Stephen Plant helped me to think through both my first and last drafts of this work as a dissertation. The dissertation upon which this manuscript is based was generously funded by the Social Sciences and Humanities Research Council of Canada.

The insightful writings of Lissa McCullough, Marie Meaney, Yoon Sook Cha, and Roberto Esposito have been incredibly helpful in my own understanding of Weil's work. I have received tremendous support from Mick Smith, Lisa Guenther, and Jackie Davies, all three of whom tirelessly gave me feedback and suggestions about how to improve this work in the areas of continental philosophy, ecology, and social justice. The ideas of this text have been informed by conversations with and support from my colleagues and friends Jacquelyn Maxwell, Tim Wright, and Josh Livingstone. The art in this collection, which embodies the ethical imperatives of Weil in the face of the climate crisis is included with thanks to Meagan Woods and Mary Mattingly for doing what they do and creating in the face of so much destruction. My love of philosophy and my continued work in academia are in many ways thanks to Antonio Calcagno, whose intellectual rigor is matched only by his kindness. And of course, my work is only possible because of the support of my family: Matthew, Mom, Dad, and Kyle. Thank you.

Abbreviations

Books and Articles by Weil

FLN — *First and Last Notebooks: Supernatural Knowledge*. Translated by Richard Rees. Eugene: Wipf and Stock Publishers, 1970.

GG — *Gravity and Grace*. Edited by Gustave Thibon and Thomas R. Nevin. Translated by Arthur Wills. Lincoln: University of Nebraska Press, 1997.

IC — *Intimations of Christianity Among the Ancient Greeks*. London: Ark Paper Backs, 1957.

IF — "The Iliad, or the Poem of Force." Translated by Mary McCarthy. In *Simone Weil: An Anthology*, 182–215. London: Penguin Books, 2005.

LP — *Lectures on Philosophy*. Translated by Hugh Price. Introduction by Peter Winch. Cambridge: Cambridge University Press, 1978.

NB — *The Notebooks of Simone Weil*. Translated by Arthur Wills. London: Routledge, 2004.

WG — *Waiting for God*. Translated by Emma Craufurd. Introduction by Leslie A. Fiedler. New York: Harper Collins, 2009.

Books about Weil

AL — Marie Cabaud Meaney. *Simone Weil's Apologetic Use of Literature: Her Christological Interpretations of Ancient Greek Texts*. Oxford: Oxford University Press, 2007.

DEB — Yoon Sook Cha. *Decreation and the Ethical Bind: Simone Weil and the Claim of the Other*. New York: Fordham University Press, 2019.

OP — Roberto Esposito. *The Origin of the Political: Hannah Arendt or Simone Weil?* Translated by Vincenzo Binetti and Gareth Williams. New York: Fordham University Press, 2017.

RP — Lissa McCullough. *The Religious Philosophy of Simone Weil: An Introduction*. New York: I.B. Tauris, 2014.

Introduction

Finding Simone Weil in an Ecological Void

This book works within a Platonic "western" ontological structure to reimagine the possibilities of that structure in light of our current ecological crisis. I do not care much for the term "western" philosophy because it suggests a division that has been proven to be contentious in myriad ways including the adaptations and import of Aristotle in Baghdad during the Islamic Golden Age when western philosophy had lost his writings altogether, or the suggestion that many of the Enlightenment-era ideas about equality and freedom were adapted from Wendat thinker Kondiaronk from New France. The "east" and the "new world" were vital contributors to this tradition but are often excluded. With that being said, it is a canon and tradition that needs to be identified, even if just to problematize that identity, and so I will be using the term "western philosophy" to imply the canonical texts taken up and taught as vitally important to the history of philosophy and the progression of certain ideas across time.[1] My decision to work within the western continental tradition and on Simone Weil in particular is in the spirit of Catherine Keller's work on theology. In her book *Face of the Deep: A Theology of Becoming*, Keller returns to the Old Testament's Book of Genesis and investigates the possibilities of the *ex nihilo* and the *deep* (*tehom*) from which creation occurs in order to eke out new feminist possibilities in the Christian tradition.[2] As Keller's eco-feminist post-structuralist reading of the Bible suggests, the dominant interpretations of a text have real effects in the world but that does not mean that they are the only, the definitive, or even the correct interpretations. Following Keller's style, I do not wish to assert *what was* but *what is possible* in the Platonic tradition through the lens of Simone Weil.

The western philosophical canon has been party to the development of particular economic, social, and political systems, which in turn have created many of our contemporary ecological ethical problems. This book assumes that the ontological and epistemic structures influencing and influenced by these social, economic, and political systems manifest in our lived experience and in particular in the lived experience of those who are excluded from or given subservient roles in society. This means that as we sit on the precipice of ecological disaster in this human-influenced geological epoch sometimes referred to as the Anthropocene,[3] we can follow the trail back from climate catastrophe and mass extinction to neoliberal

DOI: 10.4324/9781003449621-1

capitalism, globalization, and oppression of racialized and Indigenous groups, to Enlightenment-era ideals of Cartesian disinterest and radically individual subjectivity, Lockean notions of private property, a Hobbesian belief in human beings existing over and against nature, an Augustinian and Thomistic suggestion of the material sensual life as sinful, and at last all the way back to Plato's division between the shadowy and grotesque material world upon the earth and the ideal realm of the Forms and pure Good.[4] The way that we conceptualize the world, our being in that world, and the truth our knowledge asserts upon that world is the foundation of the laws we write, the policies we uphold, that which we decide is important, and that which we decide is disposable. But this is often covered over and not considered in our day-to-day lives. By the time we reach the lived reality of something like catastrophic climate change, it is difficult to imagine that this could have anything to do with an ancient Greek philosopher.

Despite the difficulties of unpacking such a link, it has been meticulously made by some scholars. For example, Jacques Derrida suggests a certain patriarchal violence in Plato's laws of opposition, which creates dichotomies still present in western thought: "soul/body, inside/outside, good/evil, seriousness/play, day/night, sun/moon, etc."[5] This book will not seek to blame Plato's thought and its adaptations across the canonical history of western philosophy and religion for the lived realities of our current ecological destruction and climate change. Rather, I will refer the reader to thinkers such as Carolyn Merchant and Genevieve Lloyd who detail such links.[6] I will make the more modest claim that traditional readings of Plato as a staunch dualist who divides the material earth from the realm of the Forms can be reimagined alongside the philosopher Simone Weil and her reading of Plato as a philosopher of interconnection and mediation. Weil offers an avenue to reimagine the divisive and apparently anti-ecological hermeneutics of Plato and some of his interpreters. She provides a philosophy that can redirect and reimagine Plato in order to maintain that which is beautiful, creative, and ethical, to acknowledge that which is violent, solipsistic, racist, and sexist, and to reimagine the possible.

Weil is searingly truthful, deceptively difficult, and as Albert Camus noted, the greatest spirit of the World War Era.[7] There are a number of reasons why approaching Weil's philosophy as an ecological ethics makes sense. Her work has an ability to hold open an absolute love of the earth alongside the existential despair that living upon the earth so often entails. This seems absolutely necessary for thinking through the terrifying prospects of extinction and destruction while still trying to live one's life. Her ability to unflinchingly note the absurdity, loss, and suffering of existence while still performing loving actions in the world is necessary for thinking through the environmental catastrophes faced by this planet. For Weil, the political and the spiritual are deeply intertwined and form a staggering ethical standard in which each person is called to sacrifice her own comfort for the love of others (albeit largely human others). Her mixture of spiritual transcendence, ethical challenge, and existential loss is prescient to our times. Her ethics demands much from us, but in the face of climate change, we must acknowledge that this is a requirement of our very survival.

Another reason I wanted to work on Weil in particular is that her work has been plagued by paternalistic editing of her ideas as well as shallow chauvinistic interpretations of her life and philosophy. In his introduction to *The Need for Roots*, T.S. Eliot states:

> Certainly she could be unfair and intemperate; certainly she committed some astonishing aberrations and exaggerations. But those immoderate affirmations which tax the patience of the reader spring not from any flaw in her intellect but from excess of temperament.[8]

Eliot flippantly casts Weil in the role of hysteric mystic. Similarly, in his introduction to *Gravity and Grace*, a compilation of both his own interpretation of Weil and reordered parts of her notebooks, Gustave Thibon wrote: "she would put the same enthusiasm and love into teaching the rudiments of arithmetic to this or that backward urchin from the village. Her thirst to cultivate minds even led to some amusing misunderstandings."[9] Thibon then goes on to give an account of Weil attempting to teach a working-class girl from Lorraine ideas from the Upanishads. More telling about his own prejudices around class and gender, Thibon finds Weil's attempts to educate to be laughable. At the very least, one could argue she acted as Socrates in *Meno*, teaching the slave boy geometry (and in the meantime, teaching Meno the nature of knowledge as recollection), but better still would be the acknowledgment that the human capability of thought transcends class boundaries and that Weil's actions exemplified a strong pedagogical commitment, not a silly flight of fancy. Both Eliot and Thibon paint Weil as strangely intelligent and headstrong to the point of foolishness. Of course, these are not traits that one would note in a male philosopher, because they would simply be expected of him. While this probably speaks more to gender norms in the 1930s and 1940s than anything else, these introductions remain in print in popular English-language editions of Weil's work and not as examples of how Weil was dismissed and underestimated based on her gender. David McLellan notes that when it comes to Weil scholarship, "facile judgements of her character have often led to equally facile judgements of her work."[10] Weil's work has often been overlooked, undermined, and misinterpreted and thus has not had the influence that I believe is possible, particularly in the realm of ecological ethics.

A Platonist far more than a Christian mystic, Weil reads Plato as a mystical rather than a purely rationalist thinker and believes that his philosophy should be understood through (not against) our material world. According to Emmanuel Gabellieri, Weil perhaps gives us the greatest alternative to traditional readings of Plato: "Simone Weil's thought is of major significance not only for contemporary philosophy but also for the history of philosophy because she offers a counter argument, and perhaps the most coherent and profound alternative, to the program of overthrowing Platonism."[11] The notion of overthrowing the Plato of dualism and the division between material earth and the ideal realm of Forms is not necessary in Weil's reading of Plato, because she sees the two realms as absolutely intertwined and the act of ethics as building bridges or realizing connections between these assumed divisions.[12] Weil's distinct reading of Plato, upon which she bases many of her own philosophical notions, offers an ontological structure conducive to ecological ethics in that it seeks interconnection and love rather than division and hierarchy.

The reading of Weil's ethics that I offer in this book is an expansion of her ethics to include not just the community of humans but also the larger non-human ecology. My distinctive mapping out of Weil's ethical thought depends upon thinking through her process of decreation. In particular, while decreation is traditionally read by Weil scholars as the act of self-abnegation alone, I have included the recreation of the self as relational as a necessary aspect of decreation. The process of decreation produces a relational self who is not egoistic but has an "ecstatic" embodied awareness of the self's very existence being entwined, even co-created, within a broader "ecology" of others. Action marks the second, creative movement of decreation. A. Rebecca Rozelle-Stone and Lucian Stone apply the helpful distinctions of (1) "the first movement of grace-Death, or the decreative event" and (2) "the second movement of grace-Recreation, or subjectivizing inspiration."[13] The first movement of decreation as a self-abnegation can bring about a problematic ethical tension because it suggests a lack of care for the self. I have conceptualized decreation as including both of these movements of grace. My reason for including both self-abnegation and action under the umbrella of decreation is twofold. First, I believe that if decreation stops at self-abnegation, it would be an undesirable form of self-indulgence and ultimately a form of destruction rather than decreation. Second, academically, I think much of the misunderstanding of Weil's decreation comes from the separation of what is meant to be a flowing movement or two sides of one coin. When decreation is separated from action, it seems like a form of self-annihilation or self-repression. But when it is interwoven with openness to other beings through a self-abnegation and a return to the site of this recognition of a relational self, then much of the argument against Weil's decreation and the personal attacks against her own life and death dissipate.[14]

Weil's ethics might be directly extended into discourses of environmental justice for human benefit,[15] but the goal of this book is to develop a more radical view of Weil as a basis for an ecological ethics in the sense of an ethics that regards the more-than-just-human environment as much more than simply a resource to be fairly distributed. In my reading of Weil, "other beings of import" can and must be expanded beyond other human beings. Expanding the Weilian sense of ethical duty beyond humans allows for a deeper understanding of our role in the Anthropocene and a reconceptualization of the human subject in relation to, and as a part of the natural world. I will establish a Weilian ecological ethics as a relational interconnection and responsibility to all other unique beings. For Weil, this is a humanist ethical project that necessitates social justice (theoretically including environmental justice in its human-focused sense), and for my own work, this ethical interconnection, responsibility, and justice extends to all beings. I believe that while this extension is not present in Weil's own writing, it is absolutely in the spirit of her thought, and I try to remain true to that spirit, if not to the letter of her writings. This is inspired in part by Lissa McCullough's suggestion that to be true to Weil's project sometimes means to read Weil against Weil in order to recognize, through Weilian ethical claims, the shortcomings in Weil's own work. For McCullough, this means a recognition of Weil's anti-Semitism as a blind spot counter to Weil's own ethical project.[16] Similarly, here I am suggesting reading Weil with Weil, deepening the transformative ethical process into a vulnerability and connection to the ecology of the surrounding earth. My expansion maintains a human priority and does not step completely into the radical equality of all living things that some ecological ethics propose. The reason for this is that in following environmental racism critiques of continental philosophy and the disdain for humanity that can arise from ecological thought in the continental tradition, I have come to understand that a radical equality among all living beings that does not prioritize human life often leads to prioritizing "nature" over the most marginalized human communities. Such a move is radically opposed to Weil's concept of ethics and my own ethical intentions for this work.

My expansion of Weil is also evident in my mapping out of her decreation as ethics. It will be obvious to anyone who has read Weil that I have mapped out a much more cohesive approach to thinking through her ideas than Weil ever did in her own writings. I have chosen to do this in order to make some of these rich ideas more accessible and clearer without flattening or changing them. As I note in the first chapter, this can be a difficult, if not impossible task, but I believe that it is necessary in my application of Weilian philosophy to ecological ethics.

The book is divided into three parts and ten chapters. The first part, including Chapters 1 through 5, sets out my reading of Weil's ethics. This map offers an ethics in which the self is abnegated in order to make space for difference and to reimagine itself relationally at the site of other beings. The first chapter serves as an apologia of sorts for my imposed mapping onto Weil's thought, which traditionally appears quite unsystematic. The second chapter introduces Weil's notion of the faculties of human knowing, loving, and willing. The third chapter examines the material embodied experiences that incapacitate these faculties and include the faculty of knowing, paralyzed by the existential recognition of absurdity; the faculty of loving, thwarted by the material necessity of absence; and the faculty of willing, impeded by the embodied necessity of suffering. The fourth chapter, on attention and mediation, explains the unique dialectic of Weil, which creates links between ideal faculty and existential necessity. This involves knowing and absurdity becoming wisdom; loving and absence becoming the love of God; and willing and suffering becoming consenting. The fifth chapter, on decreation and action, offers the concluding movement from mediation into ethical action. Wisdom acts through science, the love of God acts through art, and consent acts through work. This first part of the book is meant to give a map of Weilian ethics.

The second part of the book, Chapters 6 and 7, engages with Plato. The sixth chapter introduces Plato's *Phaedrus* dialogue and offers up the traditional patriarchal reading of Plato, which is anti-material and anti-earth. Readings of Plato's *Phaedrus* by Carolyn Merchant, Val Plumwood, and David Abram reveal the problems involved in this dualistic and anti-material interpretation of Plato. As a

Platonist herself, Weil is similarly plagued by accusations of being anti-material and destructive towards her own material body. Most of these misinterpretations of Weil hinge on a misunderstanding both of her Platonic dualism and of her key concept of decreation, the nuances of which I shall unpack throughout the book.[17] The seventh chapter offers non-dualist interpretations of Plato from Catherine Pickstock, Jennifer Rapp, and Giovanni Ferrari and culminates in Weil's reading of the Platonic dialogues, which focuses on Plato's notion of *metaxu* (μεταξύ) or the "in-between" connections at work in Plato.[18] This section establishes the dualistic reading of Plato as problematic for ecological ethics and asserts an alternative reading of Plato through Weil. Weilian Platonism will prove conducive to ecological ethics. This link to Plato is key in establishing the impact this Weilian ecological ethics can have not just on environmental thought, looking forward, but also on the history of philosophy, looking backward. Combined, the rich possibilities of past and future create a fecund ground for action in the present.

The final part of the book is a Weilian ecological ethics and includes Chapters 8, 9, and 10. The eighth chapter brings together Weil's thought and the ecological issues of the Anthropocene by unpacking the notion of the Anthropocene epoch itself and exploring theories of deep ecology. Deep ecology shares enough with my own application of Weil that it can help in unpacking Weil's concepts in an ecological framework. But while deep ecology is often criticized for not engaging enough with the social political aspects of ecological ethics, Weil's ethics keeps social political issues at the forefront and as such offers a compelling alternative to deep ecology. The ninth chapter focuses on the ways in which I think ecology in and through Weil. This involves the task of expanding which beings are ethically included in Weil's thought, and it offers counterarguments to those who read Weil as an anti-material thinker of ideal purity.

The tenth chapter offers real-world possibilities for Weilian ecological ethics. In a sense, this is an impossible task in that there is a danger of mistaking one particular example of ethical action in the Anthropocene for the whole of Weil's thought or that such examples may seem to be a drastic simplification of a larger complex philosophy. To paraphrase Eliot, my formulation of her oeuvre into examples gives a sense of pinning Weil's rich ideas to a page and watching them wriggle.[19] But I have taken care to find examples that are broad, open, and engage deeply in the possibilities of this ecological ethics. These examples serve to broaden our understanding of Weil and enact the necessity of her thought: ethical action in the material world. Rather than a conclusion, they should be seen as an invitation to creatively practice these Weilian ecological ethics in one's own life.

Notes

1 For more on Baghdad during the Islamic Golden Age, see Jim Al-Khalili, *The House of Wisdom: How Arabic Science Saved Ancient Knowledge and Gave Us the Renaissance* (London: Penguin, 2012). And for more on Kondiaronk and his influence on western Enlightenment thought, see David Graeber and David Wengrow, *The Dawn of Everything: A New History of Humanity* (New York, Picador, 2023).

2 Catherine Keller, *Face of the Deep: A Theology of Becoming* (New York: Routledge, 2003), XVIII.
3 Paul J. Crutzen, "Geology of Mankind," *Nature* 415 (2002), 23 (online: https://doi.org/10.1038/415023a).
4 See René Descartes, *Meditations on First Philosophy*, ed. Andrew Bailey, trans. Ian Johnston (Toronto: Broadview Press, 2013); Thomas Hobbes, *Leviathan* (Baltimore: Penguin Books, 1968); John Locke, *Second Treatise of Government*, ed. Dave Gowan and Chuck Greif (online: Project Gutenberg, 2003), section 27; Thomas Aquinas, *The Hackett Aquinas: Basic Works*, ed. Jeffrey Hause and Robert Pasnau (Indianapolis: Hackett, 2014); Saint Augustine, *Confessions*, trans. Henry Chadwick (Oxford: Oxford World Classics, 2008); Plato, *Plato's Republic*, trans. G.M.A. Grube (Indianapolis: Hackett, 1974).
5 Jacques Derrida, "Plato's Pharmacy," in *Dissemination*, trans. Barbara Johnson (Chicago: University of Chicago Press, 1981), 85. Also see Genevieve Lloyd, "Chapter 2: The Divided Soul: Manliness and Effeminacy," in *The Man of Reason: "Male" and "Female" in Western Philosophy*, 18–38 (Minneapolis: University of Minnesota Press, 1993).
6 Carolyn Merchant, *The Death of Nature: Women, Ecology, and the Scientific Revolution* (New York: Harper, 1990); Genevieve Lloyd, *The Man of Reason: "Male" and "Female" in Western Philosophy* (London: Methuen, 1984).
7 John M. Dunaway, "Estrangement and the Need for Roots: Prophetic Visions of the Human Condition in Albert Camus and Simone Weil," *Religion and Literature*, 17.2 *Simone Weil* (Summer, 1985), 35.
8 T.S. Eliot, "Preface," *The Need for Roots*, trans. Arthur Wills (London: Routledge, 2002), ix.
9 Gustave Thibon, "Introduction," *Gravity and Grace*, trans. Thomas R. Nevin (Lincoln: University of Nebraska Press, 1997), 6.
10 David McLellan, *Utopian Pessimist: The Life and Thought of Simone Weil* (Berkeley: Poseidon Press, 1990), 269.
11 Emmanuel Gabellieri, "Reconstructing Platonism: The Trinitarian Metaxology of Simone Weil," in *The Christian Platonism of Simone Weil*, ed. E. Jane Doering and Eric O. Springsted (Notre Dame: Notre Dame Press, 2004), 133.
12 I will unpack this further in Chapter 7 of this book, in which I set out non-dualist readings of Plato and emphasize this concept of bridge-building.
13 See A. Rebecca Rozelle-Stone and Lucian Stone, *Simone Weil and Theology* (London: Bloomsbury, 2013), 165–171.
14 For more on these personal attacks against Weil (once again, misogynistic in nature), see Chapter 5 on decreation.
15 For more on the topic of environmental justice in its many forms, see Ryan Holifield, Jayajit Chakraborty, and Gordon Walker, eds., *The Routledge Handbook of Environmental Justice* (New York: Routledge, 2020).
16 Lissa McCullough, "The Political Import of Weil's Religious Turn," at *The Self and the Selfless: Hannah Arendt and Simone Weil on Individual Action in Dark Times*, Queen's University, Kingston, ON, April 16–18, 2021.
17 Examples of this misinterpretation of Weil are examined in Chapter 5.
18 Catherine Pickstock, *After Writing: On the Liturgical Consummation of Philosophy* (Oxford: Blackwell, 1998); Jennifer R. Rapp, *Ordinary Oblivion and the Self Unmoored: Reading Plato's* Phaedrus *and Writing the Soul* (New York: Fordham

University Press, 2014); Giovanni R.F. Ferrari, *Listening to the Cicadas: A Study of Plato's* Phaedrus (Cambridge: Cambridge University Press; 1987).

19 "And when I am formulated, sprawling on a pin,/ When I am pinned and wriggling on the wall,/ Then how should I begin/ To spit out all the butt-ends of my days and ways?" T.S. Eliot, "The Love Song of J. Alfred Prufrock," in *Collected Poems, 1909–1935* (London: Faber and Faber Limited, 1936).

Part I

Growing Roots

A Reading of Simone Weil

1 Mapping an Ethics of Decreation

One of the foundational tenets for the philosophy of Simone Weil is the contradiction that God is absolutely absent from the material world and at the same time is present in the beauty of the earth, friendships, the love of the neighbor, and religious practices. For Weil, both sides of this contradiction are true in our experiences of suffering (divine absence) and love (divine presence), and she uses the contradiction between the two as the foundation of her philosophical enquiry. There is no need to "solve" the contradictions of Weil's work, because they remain unresolved in our lives, and further, they lead to a productive ethical agonism. The open-ended becoming in her ethics is at once demanding on the thoughtful individual and requires a non-egocentric orientation towards community and others. Despite their foundational purpose, the rich contradictions of Weil's work can lead her critics to read her philosophy as fragmented, inconsistent, and thoroughly resistant to the type of structure or map that I will be suggesting in this book. This chapter will set out some proposed reasons for these critiques and how Weil's contradictions are actually systematic and ethically productive rather than logically inconsistent.

Contraries in Weil's Writings and Ideas

The poetic style of Weil's writing, with its dynamism and fluidity, makes it difficult to map out her thought. Perhaps her ideas defy frameworks because her extant writings are predominantly from her notebooks and letters, which are at times only half-formulated and never fully tied together. The fragmentation of Weil's notebooks is beautifully described by Miklos Veto: "But behind the fragments, notes and sketches, behind the floating terminology and an irresistible leaning towards the extreme and the contradictory, there lie order, harmony and design."[1] Beyond her writing itself, Veto suggests that a contributing factor to the difficulty in mapping out her thought could be that while she offers concrete suggestions and ideas about our phenomenological experiences of suffering, politics, oppression, and ethics, she simultaneously gestures to the seemingly inexplicable realm of the unknown and the mystical. This pairing defies the divide that Kant established in his *Critique of Pure Reason*, in which he restricts reason to that which we experience sensibly, bracketing metaphysics (our experiences beyond the physical, namely

DOI: 10.4324/9781003449621-3

religious mystical experience) as something not accessible to reason.[2] Kant's structure of reason posits that rational systematization cannot move into or capture the complexities of the mysterious metaphysical realm. In fact, if we attempt to systematize the unknown, it will, by definition, slip through our fingers and become a flat, abstract, reified, pale-fire alternative. But Weil's philosophy contains no such exclusion of the mystery or the divine.

Weil's philosophy defies systematization because she sees the danger in the reified and absolute form of all things. This amounts to both a refusal of systematization in general and also a necessity to keep open a space for the phenomenology of mystical experience. These are two distinct but often overlapping themes in Weil's work. In her writing on Weil's apologetics, Marie Meaney posits: "Weil's approach is not systematic, but an experiential one. She starts with the experience of pain on the one hand, and her faith in God's love on the other, and strives to reconcile the two."[3] Weil's refusal to support totalizing systems in all realms of human experience is not something she maintains for the end of obtaining a mystical experience, but in her life and work, mystical influence is an outcome of this style of thought.

Despite the seemingly illusive nature of her work, careful readers of Weil all seem to agree that she does uphold a holistic overarching structure. As Lissa McCullough posits, it is "the nearly unanimous claim among her dedicated interpreters that Weil's religious thinking forms a 'coherent whole' (Springsted) or 'consistent whole' (McLellan), a 'powerfully comprehensive vision' (Williams), a sort of 'system despite the absence of any attempt to systematize' (Miłosz)."[4] McCullough goes on to note that Weil's philosophy contains a "dialectical structure that unifies the contradictions and paradoxes of Weil's religious thinking, the better to demonstrate how much consistent and thoughtful method lurks in her 'madness'" (*RP*, 11).

My project of finding and explicating inherent patterns in Weil in order to create a comprehensive map of her ethics does not change the fact that at its core, Weil's thought is anti-systematic. Weil states that philosophy is not a matter of system building but one of holding open inconsistencies. Eric Springsted explains that for Weil, "inconsistencies are not to be avoided in philosophy, and that philosophy is not a matter of system building. In the end, she also believes that thinking philosophically requires one to possess the value of detachment, and hence a readiness to be transformed."[5] But of course, Weil's style lends itself to a particular approach to reading that we could define as a certain agonistic *system* of reading. With this in mind, I will illustrate a few examples of implicit regularities in Weil's anti-systematic and anti-establishment thinking through politics, religion, and community. First, let us consider that Weil regards all political parties as totalitarian in nature. According to Roberto Esposito, in Weil's reading, a totalitarian regime is the natural outcome of the structure of democratic nation states. Unlike Hannah Arendt, who sees totalitarianism as a breakdown in the system, Weil sees it as the inevitable outcome of systematizing politics.[6] The centralization of power and the structure of upholding the party line within political organizations leads the individual to blind spots in her thinking and black-versus-white ideologies that always put one's own party in the right and the oppositional party in the wrong. Political parties do not make enough room for the political protest, discourse, flexibility, and change that are necessary for democratic action (*OP*, 5).

But for a person so bent on abolishing political parties, Weil certainly spent a large amount of time and energy working within political parties. Among her many political actions, Weil traveled to Germany in 1932 to work with Marxist activists and to Spain to fight for the Republicans in the Spanish Civil War in 1936.[7] She holds open the desire to undo the political party system *and* she tirelessly works within political parties to improve the conditions for those who are most vulnerable: soldiers and laborers, both of whom Weil saw suffering on the front lines of war and industry and both of whom she saw falling between the cracks of society. She maintains these contrary stances in both her writings and her actions. Weil does not seek to solve these contrary positions but allows them to exist side by each. As Esposito argues, "her intention is to forge the contradiction between unreconcilables to the point of reinforcing them precisely through their mutual conceptual and symbolic frictions" (*OP*, 62). This pedagogy of dialectic thinking asserts an agonism that allows for both the suffering and the love that we experience.[8]

Another example: in Weil's view, all religious organizations demand a thoughtless obedience and promote myopic thinking. In her letters to Father Perrin, a man whose spiritual generosity she admired beyond all other people's, she notes that she has detected a bias in some of his attitudes: "Notably a certain unwillingness when it comes to real facts to admit the possibility of implicit faith in particular cases."[9] She goes on to note that his slip in using the word "false" when he meant the word "nonorthodox" revealed to her an intellectual dishonesty due to his membership of the church: "It is impossible that such a thing should be pleasing to Christ, who is the Truth" (*WG*, 49). It is in this letter that Weil justifies her choice not to be baptized in the Catholic church. Most notably, she posits her search for truth above all else and her acknowledgment that sources beyond the church were often responsible for her link to the church: "Greece, Egypt, ancient India, and ancient China, the beauty of the world, the pure and authentic reflections of this beauty in art and science" are all sources of revelation for Weil and she is not prepared to close any of them off by joining one exclusionary system (*WG*, 48).

Yet Weil, unabashedly mystical in her thought, her action, and her writing, often promotes Catholic ideology. She claims a direct relation with God similar to that of Catholic saints (*WG*, 22); she praises the thought of both Saint Francis of Assisi and Saint John of the Cross (*WG*, 24); she celebrates Christianity as the religion of the slaves and counts factory workers, herself among them, as one of those slaves destined to take up Christianity (*WG*, 26); she reveals that during a recitation of George Herbert's poem *Love III*, "Christ himself came down and took possession of me" (*WG*, 27); furthermore, of great philosophical import, she finds in Christianity the harmony of the one/the Same (God as One) and the many/the Other (God as trinity), a movement that marks the assertion of contrary poles, balance, and decreation throughout her oeuvre: "The Incarnation of Christianity implies a harmonious solution of the problem of the relations between the individual and the collective. Harmony in the Pythagorean sense; the just balance of contraries" (*WG*, 33). Thus, Weil walks alongside religion without allowing herself to be taken up into its structure. Meaney notes that Weil "does not want to give unbelievers a ready-made faith, nor even encourage them to search for God. Rather, what she wishes to arouse in them is an attitude of complete intellectual honesty" (*AL*, 52). Weil uses ancient ascetic spiritual techniques in her attempts to persuade her readers to seek truth. Her highest goal is philosophy as a love of wisdom, and not religion as a love of God. This makes her religious commitments problematic to atheists and Christians alike. For the atheistic thinker, Weil's use of the term God is deeply problematic in its implications towards the dogmatic structure of religious ideology, and for the theologian, her understanding of God is somewhere between heretical and troubling depending on one's sect of Christianity. Borrowing a term from Scottish nature poet Alec Finlay, we could consider Weil's God to be "non-secular." In an examination of Finlay's work, Alice Tarbuck and Simone Kotva note that they seek "neither to sacralise nor secularise nature poetry, but instead to show how its cultivation of attentive perception is propaedeutic, rather than reductive, to the 'sacred.'"[10] This strikes me as a Weilian approach to the world and the sacred: she offers an introduction to the possibility of the sacred but her proposed conversion is simply about being attentive. What we do as we become more attentive and whether we choose to call what we attend to sacred (God) or secular (the Good) is our own decision. For Weil, maintaining her agonistic thought, the truth exists somewhere in the non-secular. Like her relationship with the political, Weil disavows *and* works within religion; she becomes an outsider on the inside. We can see an agonistic system of interpretation or "reading" emerging in Weil's life and work in which contrary poles are held open so that one can move freely between individual thinking and communal group action.

Similarly, in her reading of Plato, Weil sees society as an evil in that it controls the individual's ability to make decisions and to think towards truth/the good. For Weil, evil is understood not as a fate or a will of God, but rather as a type of karma in which a cycle of violence is perpetuated between humans (*AL*, 106). In joining the societal structure or the "great beast," the individual obstructs her own ability to thoughtfully consider the truth and as such to resist the cycle of violence. The individual becomes sloppy in her thinking and allows societal mores to make decisions for her rather than thinking them through: "The opinions of the great beast are not necessarily contrary to truth. They are formed by chance ... The trouble is that one can easily tell oneself that one is obeying God and in reality be obeying the great beast."[11] By entering the great beast that is society, the individual does not necessarily impede her own ethical truthful actions, but when she blindly follows the impulses of the great beast, she cannot be certain as to whether she acts in accordance with God and truth or as a slave to law and order. A similar distinction is noted by Hannah Arendt when she discerns the act of thinking as an ordinarily unhelpful tool in day-to-day life but the only defense against evil in dark times.[12] During historical events such as the Holocaust, when evil becomes the quotidian force of the law, it is only thinking that allows us to conceptualize beyond societal norms.[13] Each time we enter a structure or collective, we restrict our ability to think clearly and to sympathize with and love those who are outside of the structure.

And yet, Weil absolutely believes that we need interconnection and community. This can be seen in particular in the importance she bestows upon the notions of education and of roots. She suggests that education must be a cultivation of the student's ability to pay attention. This act of attending is a form of prayer for Weil. According to Angelo Caranfa, Weil's method of learning through attention involves looking but not interpretation. To pray is not to search. It is about detachment, waiting, solitude.[14] Similar to Arendt's suggestion that thinking is a death of sorts, one removes oneself from the collective and enters into individual reflection.[15] There are also major distinctions between Arendtian *thought* and Weilian *attention*: while Arendt's thinking is a removal of the I from the world of engagement into a place of seclusion and introspection with the expectation that the I will then return to the world with this knowledge, Weilian attention focuses on a removal of the I as ego-driven center of the world and an emphasis on remaining in the world while taking in its realities without totalizing them into one's own preconceived interpretation. As such, both are deaths, but for Arendt, thinking is an impermanent death of oneself in the public sphere, and for Weil, attention is a death of the ego-driven self so that the other can exist on her own terms.

The ability to learn, for Weil, is a negative effort, a self-abnegation, and most importantly a form of prayer (*WG*, 57). This act is radically individual and requires space and internal reflection, but curiously, the purpose of this practice is actually the melting away of the individual so that one can dislodge the "I" as the center of the universe and allow for the other to exist without totalizing that other. Through the communal act of education, one turns inward upon oneself in an attempt to dissolve the ego and pay attention to that which is beyond the mere individual self. From this state, one connects with God and with a deeper sense of the communal. In her book *The Need for Roots*, Weil proposes that we must cultivate a form of human community that nourishes our souls with shared recognition of beauty and mutual respect.[16] She sets out how we can reimagine society and collectivity after the Second World War and how we can establish spiritual roots within a community. One catastrophe of war is that communal roots are destroyed. As Scott Ritner notes: "The catastrophe [of World War II] is the spiritual and material crushing of the multiple rooted communities each human being needs."[17] Ritner further argues that "[i]n spite of the pessimistic qualities of *The Need for Roots*, Weil's goal is a potential postwar community oriented toward the good and comprised of individuals attentive to affliction."[18] Thus, the community structure and the thoughtful individual are simultaneously upheld despite the seeming contradiction between the two. As early as her *Lectures on Philosophy*, Weil emphasizes the interplay between individual and society and notes that "Society, it must be said, is not an aggregate of individuals; the individual is something that comes after society, who exists through society; it is society plus something else."[19] Here we can see the interconnection of individual and society, but also the priority of society *a priori* to the individual. The individual as both moral thinking person and as immoral egocentric person is made possible by the community or social contract of the collective. Only from the safety of an established social structure can the individual emerge.

Critiques and Defenses of Weil's Contradictions

In Weil, contradictions seem to abound, and this is one of the ways in which Weil's thought can become logically thick or, as Rowan Williams argues, it "collapses under a weight of contradictory pressures."[20] Williams' suggestion that Weil's work borders on "unsustainably paradoxical"[21] must be continually grappled with by Weil scholars. In the moments when Weil's seeming contradictions collide, we must carefully rethink them until we are sure that they are authentic contradictions or contraries, and not simply false or muddled thoughts. Weil establishes her agonistic methodology as follows: "as soon as one has thought upon a certain matter, [one must seek] to discover in what sense the contrary is true."[22] Nevin explains this aphorism makes clear Weilian contradiction: "These contradictory ideas are true on different planes of existence and their opposition is smoothed out on the level of supernatural love. Reason discerns the two ends of the chain, but the center which unites them is only accessible to undemonstrable intuition."[23] Thus, when one looks simply at the two poles, there appears to be a contradiction in Weil's work (for example, the contraries between the refusal to join the church and the endorsement of church doctrine). At first these two poles seem contrary, but it is through love that they are revealed as Weilian paradoxes. According to McCullough, these contradictions in Weil must be carefully thought and not simply accepted (*RP*, 40). Springsted responds to the contradictions in Weil by pointing to her use of *metaxu*, or links/interconnections, which are ontological and ethical connections between the natural and the supernatural through acts of love.[24] While I will unpack the term *metaxu* in more depth in the seventh chapter, on Weil's reading of Plato, I will note for now that the term is translated as "in-between" and is taken directly from the Platonic dialogues.[25] This use of the word implies an ethical value that may offer both a bridge (opening) and barrier (closing) between humans in communication with one another. It is significant especially in Diotima's speech in the *Symposium* when love is described as the drive "in-between" desire and beauty.[26] The structure of the human condition is herein described as a lack, a desire, and a movement towards that which could fulfill that desire. As such, the human condition is not the fulfillment of desire but rather the liminal movement towards that which we desire.

To further understand this link of *metaxu*, we must grasp the distinction between natural and supernatural in Weil's thought. McCullough defines the term "supernatural" in Weil not as a magical intervention but as

> a certitude concerning the illuminating and transformative power of pure love. Wherever love achieves a pure, unmixed presence in thought and action – and she insists that such purity is extraordinarily rare in human life – it is "supernatural" (beyond the principles of nature) in Weil's sense of the word.
>
> (*RP*, 5)

The supernatural is not magical but it is miraculous. It is not separate from the physical natural world but defies our expectation of that world. Such a definition distinguishes the supernatural both from a purely magical notion and from a purely metaphysical notion. It allows for a spiritual transcendence through a recognition of the uncontrollable mystery that Weil places beyond the world (as God or the Good). The supernatural is a miraculous act of love in the world, and we have access to this supernatural love through *metaxu*. Love is a miracle for Weil because it is the choice to act ethically towards another person when one could use violent force. For example, the lack of God here on earth in one's daily life is a natural phenomenon observable through suffering. The existence of God is a supernatural phenomenon observable through the ethical love I show towards another person when I could have harmed her. For Weil, the former is true until an encounter with *metaxu* and my ethical action in response, at which point the latter is true.

Springsted describes *metaxu* (also sometimes translated by Weil as *bridges*, although such a translation perhaps gives the word too much solidity or implies too reified a concept for the fleeting, liminal, and dynamic experiences of *metaxu*) as "things, states, or activities within this world which are subject to the correlates of good and evil ... they are things which can in a limited way be used to bring the soul into contact with God."[27] These objects are not objectively *metaxu* but become so in a combination of the object being a harbinger of the good *and* the subject's attention to said object.[28] *Metaxu* is literal, not simply metaphorical, and is neither natural nor supernatural but the interplay between the two. *Metaxu* is pivotal in understanding Weil's dialectic. According to Esposito, "Herein lies the 'key' to Weil's intentionally contradictory discourse. The very metaxy that unifies and composes is absolute division" (*OP*, 65). As the interplay between the human being and the divine good, the *metaxu* can be anything from a poem (Weil's encounter with George Herbert's poem would be such an example),[29] to a person who reveals divine love to us (most notably a friend), to a sunset (Weil speaks of nature and the beauty of the world as one of the avenues through which we come to know divinity), to a shared religious ceremony (lighting the Shabbat candles on a Friday evening), to a pair of shoes (I am thinking here of Van Gogh's peasant shoes, which famously inspired the philosophers Martin Heidegger and Jacques Derrida).[30] The examples of *metaxu* are myriad and are dependent on their relationship to the subject who experiences them.

Gabellieri argues that Weil's reading of Plato and *metaxu* offer a unique and vital modern interpretation of Plato: "Plato brings together his definitions of philosophy [the intermediary between ignorance and wisdom], love [the intermediary between the human and the divine], and dialectics [the intermediary between opinion and contemplation]."[31] Understanding *metaxu* as a pivotal relational term seems to me to be how we will keep Weil's thought from the danger of collapse that Williams suggests. The dialectic that Weil engages has no resolution and is not progress-oriented in the same way as a Marxist or Hegelian dialectic. Instead, hers is an agonistic dialectic that acknowledges the two dialectical positions without attempting to resolve them.[32] She posits the natural as the transcendental condition for the possibility of the supernatural (via *metaxu*) and as such, while one has the possibility of glimpsing the supernatural, one must always begin rooted in the material world and then open to the possibilities of the ideal Good. In this way, her agonism acknowledges the fundamental and foundational importance of our material conditions as the first order of philosophical ethical importance.

The Ethics of Balancing Contraries

Weil appreciates that ideals will always seem contradictory to the material world. She establishes a dichotomy between natural and supernatural and not between two material realities of the world. As such, the contradictions speak to the nature of desire. She insists on embracing contraries as a transcendental aspect of the human condition. They are our desire for the good. It is only through contraries that we can find *harmony*, which Weil defines as "the union of contraries" (*IC*, 168). Taking the hint from the Pythagoreans and Plato, Weil emphasizes the balance of un-synthesizable contradictions as the way to understand harmony and justice. Cha notes that for Weil, political justice is gained through equilibrium or the balancing of forces.[33] For example, ethically, Weil notes that "the contraries are myself and the other, contraries so distant that they have their unity only in God" (*IC*, 175). While my physical form may be the same genus as another person, her thoughts are always radically unknowable without the mediation of God. This mediation does not resolve the difference of the other person but allows it to exist in my ethical imperative towards that other.

For Weil, the encounter with the other person happens through the process of decreation:

> Only the renunciation of the power to think of everything in the first person, the renunciation which is not a simple transference, grants to a man the knowledge that other men are his fellows. This renunciation is the love of God, whether or not the name of God be present in the mind.
>
> (*IC*, 175)

Cha argues that Weil's ethics asserts an other whose cry is a demand. I can give that other the affirmation of subjectivity when I displace my own subjectivity in my obligation to preserve the crying other from harm (*DEB*, 27). Burns notes that it is the cry of the other that is the starting point of justice for Weil and as such, hers is a justice that is impersonal, longs for the good, and is obligated to the other. Burns posits that this is superior to a rights-based Rawlsian justice because while the rights of Rawls' liberalism are based on commercial bargaining for material goods or the indignation of another person having more than I do, for Weil, the cry of the innocent is a more primordial obligation to justice and makes me ask what I can do to help preserve the dignity of that other person.[34] Weil's ethical foundation is based upon agonism and our ability to link contradictions through *metaxu*. The ethical bind of the I to the other person is not resolved. It is a dynamic spiral rather than a line. This lack of "closure" is not a weakness in Weil's framework; rather, it is an example of Weil's dialectic in action.

Weil's writing follows the dialectic process that directs itself to the particular but also asks her reader to see the universal Good within it: through material to ideal. She guides, she acts as wayfinder, but she encourages each person to find her own way. Weil does not supply answers, she only motivates us to critically explore. McCullough posits that "In the realm of thinking, coherence is power, and Weil's thought undeniably manifests this kind of power despite the fragmented form of its expression" (*RP*, 8). Despite her aphoristic style, her early death, and her own desire to keep open well-reasoned antinomies in all areas of philosophy (social, political, religious, and ethical), the overarching coherence and whole of her work is undeniable. Weil upholds the importance of community, sharing ideas, and the transformation of the world for those most oppressed by its structures. Simultaneously, she sees the individual ability to *think* as an invaluable aspect of the human condition. Weil is interested in building or creating, but in a radically different way than reified systems. She seeks a way that does not simply mimic or invert oppressive structures. As with politics and religion, she would be cautious of any totalizing philosophical structures (for example, she finds Hegelian thought and its totalizing nature to be deeply problematic).[35] She promotes and studies an earth-bound, praxis-oriented but non-prescriptive and non-totalizing style of philosophy.

It is in this spirit that I offer not a fixed structure but an open-ended mapping out of the patterns I find in Weil's work. This is not meant to be prescriptive or absolute, but rather an exploration of what I take to be implied in Weil's writings by linking the contrary poles of our idealized human *faculties* and the existential material worldly *necessities* via our attention and our ethical *actions* in the world. These contrary poles and our attention to them constitute the process of decreation and serve to tie together Weil's overarching philosophical vision metaphysically and morally.

Linking the Natural and the Supernatural Through Decreation

Weilian contraries, according to Esposito, are a "double gaze" that is "not the contrary but the other side … [I]t is this relation that binds the two contenders in a single plot for it ties them together in the unity of a single battle" (*OP*, 49). Weil's use of contradiction can be understood in terms of the Platonic difference between the realm of the ideal Forms (the supernatural) and the material realm of earthly bodies (the natural). Weil explains the *metaxu* as both separation and bond between natural and supernatural:

> Let us imagine two prisoners, in neighbouring cells, who communicate by means of taps on the wall. The wall is what separates them, but it is also what enables them to communicate. It is the same with us and God. Every separation represents a bond.
>
> (*NB*, 497)

Weil suggests that we hold open contradictions, allow the wall to be both separation and connection, and in this act, one can witness the link (*metaxu*) between Creator (ideal) and creature (matter). This distinction between Creator (God, the

Good) and creature (human with the desire for the Good) is the contradiction par excellence for Weil in that it speaks to our human condition as desiring an ideal Good, which we can only ever glimpse. Consider the Platonic wisdom that the philosopher must be ignorant for she seeks wisdom and the lover must be in a state of want and depression for she seeks the good and the beautiful. We recognize ourselves as creatures reaching for the Creator and in so doing, we recognize contraries as two sides of one coin.

The efficacy of Weilian paradox is not held up by Weil scholars across the board, and counter to the scholars who cite a coherent system in Weil, Andrew Davison argues that Weil's mediation between creature and Creator may lack a "coherent system"; even so, he finds that it is rife with ethical value in the mediated presence of God in the material of the world.[36] According to McCullough, this dialectic between Creator and creature inaugurates the "incompatible, contradictory realities – God and world – which then gets expressed in many correlative oppositions" in Weil's work (*RP*, 15). We are creatures that long for the Creator and we may have experience of this Creator through our own decreation. Ethically and theologically, Weil's thought necessitates the difficult, even messy connection between natural and supernatural because both ethics and theology posit a desire for the good and necessitate a stance on violence and evil doings in the world. Living in the world means experiencing a certain level of violence and as such, any ethical system must seek to understand the nature of evil and violent acts. According to Weil in her account of the *Iliad*, all violence demands violence, and it is only through divine intervention that such violence can cease.[37] The paradoxical structure that seems to be in danger of collapse according to Williams or Davison may, in fact, be well supported through Weil's careful use of *metaxu*, but each individual incident of contraries must be examined by the careful reader of Weil.

Decreative Contradiction in Ecological Ethics

Decreation defies a system or codification because it is a highly personalized individuated journey, and yet Weil does give her reader an idea of what it would look like to embody the paradox of ethical existence through the act of decreation. My suggestion of an open-ended, non-rigid mapping of Weil's writings, which will be applied to an ecological ethics in the Anthropocene, is only viable if we can note that this map will never be able to produce decreation within a person. Following Weil's own style of philosophy, I acknowledge that the proposed map of decreation in this book may well have some truths to it and act as a template for understanding what it would mean to embody decreation, but simultaneously, such a structure is an abstract and oversimplified possibility of one's own winding and uncertain decreative process, which will be different in every case. And yet, with Weil's insistence that philosophy be enacted in our lives, maps, or guidelines for decreation are extremely helpful. As we have established, McCullough describes two layers of truth in Weil: the first is the fully alien supernatural that we attempt to pass into and the second is the created existence that is known through contradiction and necessity (*RP*, 37). The ethical map I will present can only ever claim to be the latter, but it aims to point towards the former. The particular care and attention that Weil gives to each concept and subsequent action is always mediated and does not necessarily suggest the ways in which that action is commonly practiced or that notion is commonly conceptualized in the world. For example, when Weil speaks of science, it is from a place highly skeptical of positivist stances on science. Hers is a science with an infusion of spiritual wisdom beyond objective truths and interwoven with both art and work. Any truly ethical action or concept will always be mediated in Weil. A science that takes only "factual" knowledge into the equation is not the science Weil suggests, as will be made clear in the chapters that follow. When this mediation process is not continually applied, it leads to misunderstandings of Weil's work.

In the remainder of this first part of this book, I will set out the structure of Weil's decreation. Weil sets out three specific faculties that she deems imperative to the human condition: *knowing*, *loving*, and *willing* (*NB*, 412). I argue that these three faculties allow human beings to give meaning and purpose to their lives. However, by the very nature of worldly existence, these faculties are paralyzed by the existential inevitability of *absurdity*, *absence*, and *suffering*. These three inevitabilities are brought on by an external factor that Weil refers to as *necessity*, and they have the potential to completely overtake the faculties by the concept of *force*. Force is the movement of worldly power that turns all human beings in its path into mere *things* or pawns. Rather than allowing one's faculties to be annihilated by these inevitabilities and thus falling into a stagnant state of existential nihilism, I will argue that Weil demonstrates how we have the possibility to simultaneously recognize both the faculty and necessity's existential barrier to that faculty: we can be attentive to these contraries shifting knowing and absurdity into *wisdom*; loving and absence into the *love of God*; willing and suffering into *consenting*. This shift does not resolve these contraries but rather allows them to remain open and offers a bridge to the actions of *science*, *art*, and *work*. Allowing for both knowing and absurdity, we utilize wisdom to practice science. Allowing for both love and absence, we shift from the love of particular objects to the love of God or the universal and we create art. Allowing for both willing and suffering, we consent to work, which is a shift to the Will of God (*FLN*, 44). This structure is not set explicitly by Weil, but the pieces of it are all present in her writing.

I will argue that this process, which gives life the possibility of ethical meaning and action, is what Weil calls *decreation*. A mimesis of God's creation of the universe, decreation is the human's removal of the small egocentric "I" that allows the individual to see herself as a member of the larger community, of humankind, or even, as I shall argue in the third part of this book, extending Weil's view, of all plants, animals: beings. In Weil's thought, if our small-ego or an inflated sense of our self as individual and radically separate subject remains, we have no possibility of moving between these contradictions and acting in a meaningful or ethical way. McCullough states that "Reality forces the 'I' to recognize that the self-centered world-construct of the 'I' must perish" (*RP*, 23). Meaney notes that it is in this vein that Weil reads Plato: "One cannot look at affliction closely unless one is willing to die to oneself for the sake of truth. This, according to Weil, is what Plato meant by '*philosopher, c'est apprendre à mourir*'" (*AL*, 163). Similar to Eastern (Buddhist and Hindu, among others) notions of self-abnegation that Weil admires and studies in various sources including the *Bhagavad Gita*, she claims that as long as we hold on to our small-I and its self-importance, we maintain a rationality that must *solve* the paradoxical (*WG*, 28). As purely logical problem-solvers, we have always already done away with the position of radically different humans. By decreating the self, the larger-sense-of-the-I can hold both the world of the small-I and the paradoxical world of the radically different person at once, and in so doing, undergo the process of ethical engagement with the surrounding world. Mapping out a structured account of Weil's philosophical thought will help us to engage more deeply in the process she calls decreation. This account risks oversimplifying decreation or abstracting from it too much, but I am balancing this risk by attending to the concrete ethical uses of a clear theoretical map or guideline for opening oneself to decreation, which I will explore in the particular context of ecological ethics.

Notes

1 See Miklos J. Veto, "Simone Weil and Suffering," *Thought: Fordham University Quarterly* 40.2 (Summer 1965), 275.
2 Immanuel Kant, *Critique of Pure Reason*, trans. and ed. Paul Guyer and Allen W. Wood (Cambridge: Cambridge University Press, 1992).
3 Marie Cabaud Meaney, *Simone Weil's Apologetic Use of Literature: Her Christological Interpretations of Ancient Greek Texts* (Oxford: Oxford University Press, 2007), 160. Henceforward, parenthetically *AL*.
4 Lissa McCullough, *The Religious Philosophy of Simone Weil: An Introduction* (New York: I.B. Tauris, 2014), 8. Henceforward, *RP* parenthetically.
5 See Simone Weil, "Some Reflections Around the Concept of Value: On Valéry's Claim that Philosophy is Poetry," trans. Eric Springsted, *Philosophical Investigations* 37.2 (April 2014), 105.
6 Roberto Esposito, *The Origin of the Political: Hannah Arendt or Simone Weil?*, trans. Vincenzo Binetti and Gareth Williams (New York: Fordham University Press, 2017), 5. Henceforth, parenthetically *OP*.
7 See Simone Pétrement, *Simone Weil: A Life*, trans. Raymond Rosenthal (New York: Pantheon Books, 1976). For her visit to Germany, see 137. For the Spanish Civil War, see 268.
8 This is by no means unique to Weil, and the politics of agonism is a lively and well-established branch of political philosophy including Giorgio Agamben, *Homo Sacer: Sovereign Power and Bare Life* (Stanford: Stanford University Press, 1998); Hannah Arendt, *The Human Condition* (Chicago: University of Chicago Press, 1998); Bonnie Honig, *Political Theory and the Displacement of Politics* (Ithaca: Cornell University Press, 1993); and Chantal Mouffe, *Agonistics: Thinking the World Politically* (New York: Verso, 2013).
9 Simone Weil, *Waiting for God*, trans. Emma Craufurd, intro. Leslie A. Fiedler (New York: Harper Collins, 2009), 48. Henceforth, *WG* parenthetically. For her comments on Father Perrin, see *WG*, 31.
10 Alice Tarbuck and Simone Kotva, "The Non-Secular Pilgrimage: Walking and Looking in Ken Cockburn and Alec Finlay's The Road North," *Critical Survey* 29.1 (2017), 33–52.

11 Simone Weil, *Intimations of Christianity Among the Ancient Greeks* (London: Ark Paper Backs, 1957), 87. Henceforth, *IC* parenthetically.
12 Hannah Arendt, "Thinking and Moral Consideration," *Social Research*, 38.3 (1971: Autumn), 446.
13 Ibid.
14 Angelo Caranfa, "The Aesthetic and the Spiritual Attitude in Learning: Lessons from Simone Weil," *The Journal of Aesthetic Education* 44.2 (2010), 64.
15 Arendt, "Thinking," 425–426.
16 Simone Weil, *The Need for Roots*, trans. Arthur Wills, intro. T.S. Eliot (New York: Routledge, 1952), 9.
17 Scott B. Ritner, "A Critique of Greatness." *Theory & Event* 26.2 (2023), 356.
18 Ibid., 360.
19 Simone Weil, *Lectures on Philosophy*, trans. Hugh Price (Cambridge: Cambridge University Press, 1978), 74. Henceforth, LP parenthetically. These lectures are recorded by Weil's student Anne Reynaud at the girl's secondary school at Roanne in 1933–1934.
20 Rowan Williams, "The Necessary Nonexistence of God," in *Simone Weil's Philosophy of Culture: Readings Toward a Divine Humanity*, ed. R.H. Bell (Cambridge: Cambridge University Press, 1993), 74.
21 Ibid., 76.
22 Simone Weil, *The Notebooks of Simone Weil*, trans. Arthur Wills (London: Routledge, 2004), 121. Henceforth, *NB* parenthetically.
23 See Simone Weil, *Gravity and Grace*, ed. Gustave Thibon and Thomas R. Nevin, trans. Arthur Wills (Lincoln: University of Nebraska Press, 1997), 156. Henceforth, *GG* parenthetically.
24 Eric O. Springsted, *Christus Mediator: Platonic Mediation in the Thought of Simone Weil* (Chico: Scholars Press, 1983). For an extended look at *metaxu*, see Chapters 4 and 7.
25 The term *metaxu* is used in Plato's *Symposium* and *Phaedrus*. Plato, *Symposium*, ed. M.C. Howatson and Frisbee C.C. Sheffield, trans. Howatson (Cambridge: Cambridge University Press, 2008). Plato, *Phaedrus*, trans. Benjamin Jowett (Bulgaria: Demetra Publishing, 1870).
26 See Plato, *Symposium*, 202a3, p. 37.
27 Springsted, *Christus Mediator*, 198.
28 Peter Winch notes that Springsted pushes *metaxu* beyond mere metaphor, and Winch uses this interpretation but shifts to perspectival reading rather than Springsted's religious reading. See Peter Winch, *Simone Weil: "The Just Balance"* (Cambridge: Cambridge University Press, 1989), 81.
29 See *WG*, 27.
30 Martin Heidegger, "The Origin of the Work of Art," in *Basic Writings* (New York: Harper Collins, 2008), 159; Jacques Derrida, "Restitutions of the Truth in Pointing," in *The Truth in Painting*, trans. G. Bennington and I. McLeod (Chicago: University of Chicago Press, 1978).
31 Emmanuel Gabellieri, "Reconstructing Platonism: The Trinitarian Metaxology of Simone Weil," in *The Christian Platonism of Simone Weil*, ed. E. Jane Doering and Eric O. Springsted (Notre Dame: Notre Dame Press, 2004), 146.
32 This dialectic will be set out in detail through the next three chapters as we observe the faculties of knowing, loving, and willing, as opposed to the contrary certainties of absurdity, absence, and suffering.
33 Yoon Sook Cha, *Decreation and the Ethical Bind: Simone Weil and the Claim of the Other* (New York: Fordham University Press, 2019), 49. Henceforth, *DEB* parenthetically.

34 Steven Burns, "Justice and Impersonality: Simone Weil on Rights and Obligations," *Laval Théologique et Philosophique* 49.3 (1993), 477–486.

35 Simone Weil, "Pre-War Notebook," in *First and Last Notebooks: Supernatural Knowledge*, trans. Richard Rees (Eugene: Wipf and Stock Publishers, 1970), 17. Henceforth, *FLN* parenthetically.

36 Andrew Davison, "The Mediating Possibilities of Absence in the Thought of Simone Weil," *Theology* CXII.865 (January/February 2009), 10–11.

37 Simone Weil, "The Iliad, or the Poem of Force," trans. Mary McCarthy, in *Simone Weil: An Anthology* (London: Penguin Books, 2005), 206. Henceforth, *IF* parenthetically.

2 The Faculties

Weil never fully unpacks why she has chosen the three particular faculties of knowing, loving, and willing as representative of the human condition, although she describes them as the routes by which we come to recognize the sheer impossibility of absolute meaning in our lives:

> We are beings which know intellectually, exert our will, and love; and as soon as we bring our attention to bear on the objects of our knowledge, our will, and our love we are manifestly bound to recognize that there are not any of them which are not *impossible*. It is only lying which can conceal this fact from us. The consciousness of this impossibility forces us continually to desire to seize the unattainable through and beyond everything we desire, know and want with our will.
>
> (*NB*, 412)

In other words, Weil posits we are forced to realize that even as we exercise these faculties, they are tempered by the impossibility of delivering on their implicit promise of foundational meaning. But before turning to this impossibility, it is wise to consider the promise of meaning that Weil sees as implicit in these three faculties. This will have to be my own interpretation of what Weil might have seen as being ascribed to these faculties, because she engages only in their impossibility. But of course, to argue this impossibility is to acknowledge the human desire for these faculties to deliver an absolute meaning. For Weil, each faculty offers an *ersatz* form of meaning or an incomplete substitute for meaning and as such I will describe the faculties as promising what Weil will ultimately posit as an illusion. The faculties can only gain true meaning when they have gone through her proposed process of mediation. This is a necessary journey from innocence to experience that allows for ethical action.

According to psychiatrist and Holocaust survivor Viktor Frankl, the search for meaning is pivotal to human beings and necessary for the possibility of human fulfillment.[1] Pushing back against Freud's suggestion that humans are driven by a desire for pleasure, Frankl argues that in fact, *meaning* is the foundational desire and drive of humanity. While Weil does not emphasize *meaning* explicitly, she

DOI: 10.4324/9781003449621-4

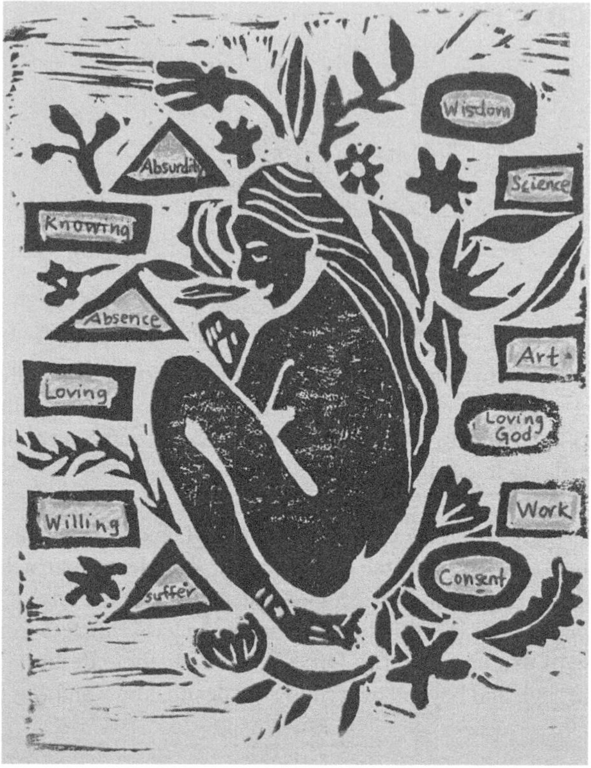

Figure 2.1 Kathryn Lawson, *Decreation.* Linocut and ink (2023).
A diagrammatic mapping out of Weil's thought to assist in conceptualizing the first part of the book.

does place a strong emphasis on *truth* as a foundational tenet for living one's life, and the two are interconnected points. Truth can be posited as a possibility of the faculties and a justification of their meaning making. In my reading of Weil, the faculties of knowing, loving, and willing are three different avenues that can offer a foundation for meaning in the world and as such offer us a sense of control and truth. Exercising these faculties, humans create meaning; we create human worlds that promise a way to ground an otherwise tumultuous and uncertain existence. But for Weil, as I shall unpack in the next chapter, because these are faculties pursued through human will, they are ultimately *ersatz* or limited in their ability to provide meaning. This is an example of her dialectic thinking, in which she takes the foundations we may assume to be true and then she considers the opposite, allowing for a more nuanced picture of that which is true. But before we can move into the dialectic, we must understand what is being expressed in a positive way through the assertion of the faculty. To this end, I will examine *knowing* through David Hume and the Vienna Circle, *loving* through Martin Heidegger and bell hooks, and *willing* through Friedrich Nietzsche and René Descartes. These philosophers offer one example of how the faculty could offer us meaning and as such, what we have to lose if said faculty is overturned. To be clear, I am not asserting that these thinkers are representing *ersatz* meaning. Indeed, to use Weil's terminology, most of them address the limits of the faculty within their own work. They are often acutely aware of the limitations that Weil also suggests. I will work with each of them briefly to present the potential of the faculty but will not go in depth as to how this informs their larger oeuvre or how they engage with the faculty's limitations. I find it helpful to turn to these thinkers because Weil has not set out what exactly she intends by these faculties in their "possible" and "pre-mediated" forms and as such, this is my own possible answer to the meaning implicit in the faculties.

The Faculty of Knowing

We can imagine, as Weil suggests, that in the search for a foundational meaning for human existence we might call upon different frames of reference closely aligned with the faculties. The first potential answer is a knowledge that offers a consistent framework of well-established facts and theories for understanding existence. In order to establish the meaning that may be posited by the faculty of knowing, I will turn to the philosophical example of the Scottish philosopher Hume (1711–1776) as he was adapted by the Vienna Circle. The faculty of knowing, as I will set it out in the Vienna Circle, is distinct from Weil's own understanding of knowledge, as she strongly disagrees with the positivist interpretation of science. Instead, it is a way that this *ersatz* faculty can make meaning.

Hume offers a strong epistemic framework for the faculty of knowing. An empiricist, he posits that any claim to knowledge must originate in sense experience.[2] In many ways a product of the Scientific Revolution of the seventeenth century, Hume noted that theory needed to be reinforced and supplemented by experimentation and empirical sensory data. For Hume, our sensory impressions of the surrounding world provide the material necessary for our ideas, or for true knowledge. Without a sensory impression for support, an idea cannot be the basis of truth. These *matters of fact* based on sensory impressions relate to one another to provide a foundation for higher-order knowledge, such as mathematics. The *relations of ideas* are still founded on empirical sensory impressions, but not directly from impressions in the way that he derives matters of fact.

Hume's two categories of *matters of fact* and *relations of ideas* make up the entirety of possible knowledge. Any claims that are not founded on matters of fact or relations of ideas are nonsensical and meaningless in the Humean epistemology. As such, Hume offers the distinction between *fact* and *value*, or, that which is objectively verifiable and that which is merely subjective. Human knowledge only extends to matters of fact or relations of ideas. Everything else is nonsense and beyond the scope of verifiable epistemic truth. This general outline of Hume's epistemology can offer one possible basis for the faculty of knowing as the ability to distinguish verifiable objective fact from individual subjective value statements or opinions. Knowledge can be herein determined through this process of distinguishing objective matters of fact and relations of ideas (true knowledge) from subjective value (mere opinion).

Beginning with a Humean epistemology, the Vienna Circle of the early twentieth century (c. 1922–1936) shifted their focus to propositions. This group of logical positivists argued in favor of two types of propositions based upon Hume's matters of fact and relations of ideas. For a proposition to be empirically demonstrable, it must be a matter of fact, which can be proven by empirical data, or a relation of ideas, which is true by definition (e.g., all widows are women whose husbands have died) or proven through a higher-order verification such as a mathematical proof. Notably, both true and false matters of fact and relations of ideas are meaningful for the Vienna Circle because both offer a truth claim. However, all notions without logical empirical verification are designated as meaningless in this epistemology. Thus, all propositions that can be verified, even if they are incorrect, have epistemic value. For example, a spring day may appear to be warm and thus one dresses in a light jacket, but it is actually bitterly cold. Just because one improperly assesses the data does not make it meaningless. It can still be verified, and the matter of fact adjusted accordingly. In the case of the spring day, one returns to the house for a coat and hat. The upshot is that we have established the potential of knowledge to bring about meaning through empirical truth. As such, it gives an idea of the meaning that could be lost if this faculty is called into question, as it will be in the following chapter's section on absurdity.

As with all of the potential foundations of meaning, there are many forms of understanding that can arise from an empiricist epistemology. Hume and the Vienna Circle represent just two, albeit exemplary, cases. From the faculty of knowing, we are given meaning that relies on the fact of empirically verified logical truths upon which we can build our meaning and our world. Broadly speaking, the faculty of knowing promises a world based upon logical empirical truths and an unshakable scientific certainty.

The Faculty of Loving

In contrast to those who seek meaning in the store of empirically based knowledge they acquire, others will point to relationships as the keystone of a meaningful life and a foundation for our existence. Our world is meaningful because of the beings and things that we share it with. Our emotional connection to other beings colors our existence and makes our lives worth living. This second mode of meaning making springs from the faculty of loving: love for people, love for places, love for things. I will turn to Heidegger to recognize this faculty at work in relation to people, places, and things, and then to bell hooks to recognize this faculty in relation specifically to particular other humans and radical social political change. As in the previous section's foray into Hume and the Vienna Circle, while Weil deals with love in many ways, she does not give a strong account of love as an *ersatz* faculty and as such, Heidegger and bell hooks offer two exemplary ways to think through the faculty of loving and the meaning that such a faculty can bestow. Both Heidegger and bell hooks note the limitations (void, death, etc.) of love and Care

(*Sorge*) in their own works, although I will predominately explore their work as a means to explain the faculty and its potential for making meaning.

In *Being and Time*, Heidegger sets out Care (*Sorge*) as the being of *Dasein* (or "there-being," which is Heidegger's term for the human existent beyond the baggage of the traditional philosophical human "subject").[3] Heidegger's use of the term Care refers to an ontological way of being, or an attunement of Being-in-the-world that is more fundamental than caring relationships between particular existents. It is a Care that is based on an involvement in and an awareness of the surrounding world: "a constitution of being which always already is taken as a basis. This constitution first makes it ontologically possible that this being can be addressed ontically as *cura*."[4] Here, *cura* refers to the ontic level of care between particular people, whereas *Sorge* refers to the ontological level of Care as a way of being. Heidegger is concerned with ontological Care (*Sorge*) and not ontic care (*cura*). Ontologically, we organize our Being-in-the-world through Care and "Reality is referred back to the phenomenon of care."[5] For Heidegger, this notion of Care is a technical term that involves a human's being *thrown* into the world without any choice, namely being born into a particular body and life without consenting to that existence (*facticity*); having *projects* (one's own-most possibilities) that can create a meaning, give a sense of control, or a ground to one's existence; and *falling prey*, which is an existential mode of being-in-the-world that favors the everyday framework of the masses (*das Man*), which sees humanity as something separate from the world, covering over the authentic reality of Dasein as entangled in the world with others.[6] Again, we can herein note Heidegger's strong awareness of our human likelihood to engage in *ersatz* variations of the faculty, to use Weil's language.

These elements of Care are revealed in Dasein's ultimate possibility, namely, death.[7] A person's own death is the most extreme and unavoidable reality of existence and reveals the limit of Dasein's projects in the sense that her death is the end of all of her goals and projects. Anticipation of Dasein's own death gives her projects meaning and reveals that Dasein has always already fallen prey to inauthentic being. A person feels a sense of *anxiety* at the fact that she is given no reason or ground for being other than what she herself establishes and makes meaningful despite the fact that she is born into a world over which she has no control.[8] Her thrownness into the world encourages inauthenticity in the sense that it makes it easy not to consider our own inevitable death and rather to take for granted the inauthentic meaning structures of society. But in moments of authenticity, she anticipates her own death as the end of that being-in-the-world.[9] To authentically Care involves an anticipation of her own death and a resoluteness in the face of the guilt or responsibility one feels for being the ground or meaning of her own being.[10] This type of ontological Care for the self as always already in the world with others and concerned about one's own projects in and with the world/others suggests a form of loving quite distinct from our ontic colloquial use of the term care or love. For Heidegger, "Meaning signifies that upon which the primary project is projected, that in terms of which something can be conceived in its possibility as what it is."[11] To have meaning is to be capable of having something *matter* to you as a relational being-in-the-world who is thrown, project-oriented, and falling prey to inauthenticity, and this is brought about through Care.

I here include Heidegger's ontological Care as an example of the existential belonging to the world that I see as necessary for a full understanding of what Weil might intend with the faculty of loving. Heidegger's ontological and impersonal notion of Care is distinct from a narrower understanding of love for particular others. The ontological aspect of love expands love beyond an interpersonal or romantic notion. The purpose of these faculties is the creation of meaning, and this is precisely Heidegger's motivation in theorizing *Sorge* as a (relational) creation of the meaning of being. The hunter has Care for her gun and the rabbit she stalks just as much as the father has Care for the flour and sugar he uses to bake a cake for the child for whom he also Cares. This ontological structure of Care asserts a way of Loving that gives each action meaning and purpose. The faculty of loving, as I would like to conceptualize it for the purposes of understanding Weil's philosophy, also includes an ontic or ethics-based interpersonal notion of love, which I will now unpack through the work of bell hooks.

In Black feminist thought, love has been taken as a form of radical resistance by thinkers such as Anna J. Cooper, Traci West, Chela Sandoval, and Patricia Hill Collins. Love is an ethically challenging proposition because it requires vulnerability and the recognition of interconnection. Love is not simply a peaceful, beautiful, easy thing. To love with the full recognition of absence and loss is one of the most challenging things a human can endeavor to take upon herself. In this, we see that bell hooks, too, rejects love as a simplistic *ersatz* faculty. The act of love and the recognition of interconnection does not mean that one accepts violence, abuse, or ethically questionable actions. Instead, one holds the other accountable and simultaneously recognizes that the other is connected to oneself. A theorist of the radical possibilities of love, bell hooks notes:

> Sadly, accepting human variety means that we must also find a way to positively connect with folks who express prejudicial feeling, even hatred. Committed to building community we are called by a covenant of love to extend fellowship even when we confront rejection. We are not called to make peace with abuse but we are called to be peacemakers.[12]

According to bell hooks, an ethic of love is necessary to counter an ethic of domination. In cultivating love rather than perpetuating or inverting systems of oppression, we cultivate a radical new possibility for society and community. Generally, people fight against one system of domination in favor of another system of domination, but an ethic of love has the potential to break this cycle. Just as all political parties have the natural end of totalitarianism for Weil, bell hooks notes the ethics of domination at play in both the system in power and the system attempting to overthrow that power:

> This is why we desperately need an ethic of love to intervene in our self-centered longing for change. Fundamentally, if we are only committed to an

> improvement in that politic of domination that we feel leads directly to our individual exploitation or oppression, we not only remain attached to the status quo but act in complicity with it, nurturing and maintaining those very systems of domination.[13]

Love is not simply love. It is radical, difficult, and steeped in social justice. This type of love offers a meaning that is embedded in ethical care for other people, and it is exemplary of the meaning that is possible from the faculty of loving.

Through the faculty of love, the foundation of the world rests on the fact that we care for the beings in it both on an ontological level and an ontic ethical level. This can be perverted or morphed by our biases and the challenges of the worldly situation into which we are thrown and our lived experiences. For example, unrequited, abused, or neglected love becomes anger, revenge, self-absorption, obsession, or any number of motivations that may *seem* to be the opposite of love. On the other hand, love can manifest as kindness, creation, or any number of care-filled emotions. It can be through love or love's lack that meaning is played out in the world. The faculty of love can offer meaning in many ways, but the ethical social-justice-oriented way of bell hooks and the ontological inescapability of concerned involvement of Heidegger are two exemplary cases that can assist in conceptualizing what is at stake if this faculty falters in its meaning making.

The Faculty of Willing

Third, and finally, there is the foundation of meaning that we make through our willpower. With hard work and perseverance, we build shelter, harvest food, craft religious icons, erect community centers, and assert our place on the earth through a will that creates worlds. In order to conceptualize this faculty, I turn to the notion of will in the philosophers Nietzsche and Descartes, both of whom have an exemplary understanding of the will as a faculty of meaning making. As with the last two faculties, these are by no means extensive investigations of these philosophers, but rather, they offer insights into how we might more fully conceptualize what Weil may have seen as at stake in establishing the will as a maker of meaning. And as with knowing and loving, the explanation of willing is not a reflection upon a Weilian understanding of the will, but rather establishes the will as a possible meaning maker that Weil will consider and ultimately challenge through the process of mediation.

Nietzsche's notion of the *will to power* and the philosophy of the *Übermensch*, whom a human can will herself to become, exemplifies this mode of meaning making. This process is seen in the character Zarathustra in Nietzsche's text *Thus Spoke Zarathustra.*[14] Throughout this text, Nietzsche follows the progress of a pilgrim-cum-guru on a spiritual journey to self-actualization in a process of overcoming his own humanity through his will to power. The tale follows Zarathustra's metamorphosis from camel, a reverent beast who carries the heavy burden of existence; to lion, asserting a will that overcomes the burdens demanded by society; to child, who creates the world anew.[15]

With her willpower, the human takes control, asserting the I as the ultimate creative force of existence. For Nietzsche, "man is something that must be overcome"[16] through a gradual recognition of the will to power and the emergence of the *Übermensch* (the over-human, or beyond-human). In this movement beyond herself, the aspirant recognizes her own willpower as capable of forming her own reality. Zarathustra's recognition of his own willpower launches him into the beyond-human. In this way, the will makes a god of a human by situating the power to create the world in individual human willpower. This power of the will is described by Zarathustra as a brute force: "Yes, something invulnerable, unburiable is within me, something that rends rocks: it is called *my Will*."[17] Beyond this forceful power, will is also something beautiful and creative to Zarathustra: "Where is beauty? Where I *have to will* with all my will; where I want to love and perish, that an image may not remain merely an image."[18] Will takes an idea and creates action and life. It molds the very world. As such, it is beyond the cycle of existence as suffering. It can transcend suffering and bring about both freedom and joy: "All *feeling* suffers in me and is in prison: but my *willing* always comes to me as my liberator and bringer of joy. Willing liberates: that is the true doctrine of will and freedom – thus Zarathustra teaches you."[19] Freedom, joy, and force are wrought by the individual who wills herself into the *Übermensch*. As such, will is the faculty par excellence for Nietzsche, although he wouldn't use those terms. It allows humans to overcome the expectations, drudgery, and suffering of existence. Ultimately, it allows one to take up the God-like position of making her own meaning in the world.

While Nietzsche offers one of the best examples of finding meaning in the faculty of willing, another fine example comes from Descartes, whom Weil studied closely, and indeed, was even the topic of her dissertation at the École Normale Supérieure.[20] In the fourth of his *Meditations*, Descartes considers the distinction of truth and falsity. Explaining how humans fall into error, Descartes compares the faculty of understanding and the faculty of willing. He asserts the faculty of will as that which makes humans like God because it is, for Descartes, the only limitless faculty that humans possess: "It is only my will or my freedom to choose which I experience as so great in me that I do not apprehend the idea of anything greater. Thus, through my will, more than through anything else, I understand that I bear a certain image of and resemblance to God."[21] In the will, Descartes observes a limitless freedom, and he argues that no other human faculties are as "perfect" and expansive as the will.[22] Indeed, the will is so powerful compared to the limited scope of the faculty of understanding that the imbalance between the two is what causes errors in human knowledge. For Descartes, while the will offers the human a closeness to God, it is quite simple in that it is "only a single thing and is, so to speak, indivisible."[23] The will operates as an all-or-nothing faculty: we either assent or refuse to assent. Without the freedom to assent or refuse assent we have no freedom of will at all. An unfree will is not will. As such, while the imbalance between understanding as limited and willing as limitless may cause errors of judgment, it is better than having no will at all. It is better insofar as it is the ground of agency and something for which we can thank our creator in spite of the error and sin for which it allows. This likeness to God that we discover in our will offers us a meaningful position in the great chain of being: "I am, as it were, something intermediate between God and nothingness."[24] The Cartesian will gives humans meaning on two counts: as the faculty that is most God-like and as such, the faculty that provides intimations of the divine; and as the faculty that we can choose to align with understanding and thus avoid errors of judgment, indeed avoid sin. As such, the faculty of willing provides spiritual as well as epistemic meaning for Descartes.

The faculty of will in Nietzsche and Descartes creates meaning through a recognition of the human's god-like potential. Harnessing this faculty can center the human either as a god of sorts or as closer to God in the great chain of being. Of course, these are only two ways that the will may offer meaning, but both are precisely the type of meaning that Weil will problematize through her dialectic process. These three faculties of knowing, loving, and willing are not exhaustive in pinpointing the ways in which we create meaning, but they are extensive in their ability to provide an ethical-ontological map of different modes of meaning making, and I think that this is why Weil chose them: they represent the foundations of our human worlds and as such, they embody all that crumbles if (and according to Weil, *when*) their promise to provide meaning falters. Taken as the basis of one's meaning in the world, they seem to have the potential to form the epistemic boundaries of our worldly experience, the metaphysical and ontological structures of our being-in-the-world, and the ethical duties that we owe to other worldly beings. We can imagine that this may be why Weil chooses them as the beginning point for considering the necessity for ethics and the existential impossibility of absolute meaning.

Notes

1 Viktor E. Frankl, *Man's Search for Meaning* (Boston: Beacon Press, 2006).
2 For Hume's empirical epistemology, see David Hume, *A Treatise of Human Nature: Being An Attempt to Introduce the Experimental Method of Reasoning into Moral Subjects* (Oxford: Oxford University Press, 1978).
3 Heidegger claims that Dasein is something more-than or beyond our concept of the human being. Martin Heidegger, "Care as the Being of Dasein," division one, Chapter 6, *Being and Time*, trans. Joan Stambaugh, ed. Dennis J Schmidt (Albany: SUNY Press, 2010), 175–220. When I use "Care" I refer to the ontological *Sorge* and when I use "care" I refer to the ontic *cura*.
4 Ibid., 192.
5 Ibid., 203.
6 Ibid., 272 and 198–199.
7 Ibid., 248.
8 Ibid., 284–285.
9 Ibid., 289.
10 Ibid.
11 Ibid., 309.
12 bell hooks, *Belonging: A Culture of Place* (Abingdon: Routledge, 2008), 67.

13 bell hooks, "Love as the practice of freedom," in *Outlaw Culture: Resisting Representations* (Abingdon: Routledge, 2015), 290.
14 Friedrich Nietzsche, *Thus Spoke Zarathustra*, trans. R.J. Hollingdale (New York: Penguin Books, 2003).
15 Ibid., 54–55.
16 Ibid., 83.
17 Ibid., 135.
18 Ibid., 145.
19 Ibid., 111.
20 Simone Weil, "Science et Perfection dans Descartes," in *Sur la Science* (Paris: Gallimard, 1966).
21 René Descartes, "Fourth Meditation," section 57, in *Meditations on First Philosophy*, ed. Andrew Bailey, trans. Ian Johnston (Toronto: Broadview Press, 2013), 68.
22 Ibid.
23 Ibid., 70.
24 Ibid., 66.

3 The Power of Force

While Weil is greatly indebted to the Greeks, European philosophy, and Christianity, she also draws heavily upon the *Bhagavad Gita* and Eastern philosophy in general (Buddhist and Hindu), particularly in her concept of decreation.[1] The illusion of individuality (an illusion dispelled in the process of decreation) causes us to think of meaning and purpose as being created though the triumph of our faculties of knowing, willing, and loving. The *Bhagavad Gita*, a text closely studied by Weil, says:

> Actions are all effected
> By the qualities of nature;
> But deluded by individuality,
> The self thinks, "I am the actor."
> When he can discriminate
> The actions of nature's qualities
> And think, "the qualities depend
> On the forces of nature," he is detached.[2]

This text argues that we are deluded if we believe that we are solely responsible for the outcomes of our actions. This is to forget the force of nature and the necessities of our existence. The process of decreation is a recognition of the impossibility of solipsistic totalized control when faced with *forces of nature*, as the Gita calls them in the above quote. Herein, we recognize that the faculties cannot possibly deliver on a promise of providing unshakable meaning. This recognition is a key element of what Weil calls decreation and, in part, what the Gita calls detachment. Weil claims that "as soon as we bring our attention to bear on the objects of our knowledge, our will, and our love, we are manifestly bound to recognize that there are not any of them which are not *impossible*" (*NB*, 412). This is why I now turn to a more detailed analysis of what it is, for Weil, that makes these faculties impossible by unpacking the concept of *force* in conjunction with Weilian *necessity* and offering an ecological example of force and necessity in the Anthropocene. I will then note an ethical critique of Weil's force and defend her ethics via her notion of

DOI: 10.4324/9781003449621-5

balancing force. Finally, I will turn to how force compromises the three faculties through the necessity of absurdity, absence, and suffering.

Defining Force

In her essay on the *Iliad*, Weil argues that the lead character of Homer's epic is force itself (*IF*, 183). Weil maintains that force has always been at the very center of human history and in making force the hero of the poem, the *Iliad* becomes one of the greatest artistic works of all time. What Weil refers to as force is not an *élan vital* or an autonomous life force, or any kind of "evil" working on matter, but a quality of matter itself. It cannot exist separately from the material world and is perhaps best conceptualized as the energy in matter. Despite its importance in her philosophy, force defies a straightforward definition. Weil instead uses an undefined *x* to describe force.

According to historians, it is René Descartes who initially inserts *x* as the marker of the unknown in his 1637 treatise on geometry and thus, perhaps unwittingly, thrusts the letter *x* forward as the symbolic representation of all that we seek and do not know.[3] As both an avid mathematician with a focus on geometry and a close reader of Descartes, Weil adopts this algebraic symbol as the representation of force. She asserts that force

> is that *x* that turns anybody who is subjected to it into a *thing*. Exercised to the limit, it turns man into a thing in the most literal sense: it makes a corpse out of him. Somebody was here, in the next minutes there is nobody here at all; this is a spectacle the *Iliad* never wearies of showing us.
>
> (*IF*, 183)

Thus, force is largely irresistible and often unknowable, but the effects of force are clear: the annihilation of meaning, purpose, foundation; the transformation of a subject into an object.

It is a common mistake to attempt to control force, as it is not something humans can control through willpower. Force controls us, and it is a fool who thinks that she can assert absolute control over force when she is caught up in its throes:

> Force is as pitiless to the man who possesses it, or thinks he does, as it is to its victims; the second it crushes, the first it intoxicates. The truth is, nobody really possesses it. The human race is not divided up, in the *Iliad*, into conquered persons, slaves, suppliants, on the one hand, and conquerors and chiefs on the other.
>
> (*IF*, 191)

Even those who wield force become objects and lose the subjectivity that makes them thoughtful individuals. As Yoon Sook Cha notes, force is "a complex field of material, physical, and psychic conditions and processes that constitutes the scene of vulnerability and frames the response to the suffering born of this exposure" (*DEB*, 11). Force is the situational space in which we are called to ethical action.

At the scene of vulnerability, we cannot control force, but we can choose not to be the tool through which it objectifies another human being and in so doing, we can bring it into balance.

Force and Necessity

Following basic physics, we know that matter is energy, and in conceptualizing Weil's force, we could say then that force is the energy in matter, which expresses power in ways that affect and are affected by human actions. Force is always largely unknowable and does not adhere to expectations. *Necessity*, on the other hand, is that which governs a mathematical equation. Necessity *can* be understood, and through it, one can recognize beauty and the good. Force seeks out power. It does not recognize that which is good and beautiful. Yet force is not completely beyond human control, because necessity implements limitations on force in the very physical laws of material existence: "Necessity always appears to us as an ensemble of laws of variation, determined by fixed relationships and invariants. Reality for the human mind is contact with necessity" (*IC*, 178).

While no one can strong-arm force, it is possible to bring force into an ethical balance. Indeed, force maintains an ethical saving grace: it brings about the kind of suffering that is a precondition of truly understanding justice and love. The affliction of oppression creates the conditions for the possibility of love. To suffer creates the possibility within a person to choose not to inflict suffering upon another. This choice halts the cycle of violence through an ethical balance that can limit force before we are taken up in its sway. With one's own balanced reaction attending to the realities of suffering and ultimately love, force's power is curtailed. By observing the physical necessity that limits force, one can recognize the potential construction of material and ethical limitations to force. Force admits to limits, and we have an obligation to bring it into an ethical balance between suffering and love.

Necessity denotes the relation between a human's material bond to the natural world and the small impersonal part of her which longs for the supernatural Good. As discussed in the first chapter, this distinction between the natural and supernatural in Weil can be understood as the difference between the Platonic realm of the Forms (the supernatural) and the desire within the individual to attain the ideal Good from their position in the material realm (the natural). *Necessity* governs the mediation between "natural" desire and "supernatural" Good, whereas *force* governs only the desires of the "natural" world. The Weilian natural refers here to the material patterning and expression of the universe and not environmental philosophy's distinction of "nature" as opposed to human artifice. For Weil, humans have an ability to transcend in terms of recognizing a "supernatural" aim, which is directed towards God or the Platonic Good. Force is purely natural and does not engage with the supernatural. For Weil, force, in human terms, is not a matter of desiring the Good but of desiring more and more power within the laws of physics and "human nature." Necessity, by contrast, mediates between natural and supernatural by dictating the parameters of the possible within earthly reality. Both force and necessity are qualities of matter and as such, they are different in degree rather than being radically different in kind: "This force, which governs the world and makes every man obey, this force is the same as that which the human mind conceives of under the name of necessity" (*IC*, 182). Force is limited in its scope by necessity, or the empirically real limitation of our existence.

Force and Necessity in the Anthropocene

In the Anthropocene, predominant modern social forms have become characterized by and obsessed with possessing force and exercising it against the "natural world." The desire for more and more power, for literally more and more energy in the form of natural gas, oil, bitumen, etc., has made some humans believe that force will always be on our side and furthermore that we are the ones controlling force. But necessity is revealing the limits of force through the heat domes, wildfires, droughts, tornados, melting glaciers, species extinction, etc. brought on by global heating. Necessity reveals that humans are not in control of force and that humans are not separate from nature; we are part of, and party to, the patterning of force. In the process of making nature an object, the human also becomes a mere object or tool for force's acquisition of more power, more energy.

Rather than heeding these limits of necessity, some humans turn to potential technological reversals of climate heating or geoengineering projects that could allow for limitless energy consumption.[4] These solutions ignore necessity and focus on maintaining force. They are solutions that maintain the human centrality of the issues at hand rather than turning to a more geological approach, one which recognizes humans as unable to wield force indefinitely.

But consider that for Weil, it is precisely at this moment, when one has force on one's side and chooses not to use it against another person (or in my adaptation of Weil, another being or thing), that one can bring balance and act ethically. It is in this moment that the potential to expand the understanding of the human "self" as relational and interconnected to the surrounding world becomes possible. Here, one can bear witness to necessity and recognize the ethical wrongdoing of thoughtlessly wielding force and therein objectifying beings (both human and non-human), vegetation, and the things of the earth. One recognizes the need for a new framework beyond the Anthropocene, a paradigm of love not of force. Weil pushes back against the notion of will in particular because it exemplifies a thoughtless adherence to force and a hunger for power and ego-centrality. These ecological issues offer an apt example of the difference between force and necessity while revealing how this distinction brings about the possibility for ethical action.

An Ethical Critique of Weil's Force

The great equalizing power of force assumes that the subjugated and the subjugator are controlled by the same mechanism and as such both are equally blameworthy or innocent. If we conceive of force as making all people objects, we lose a robust sense of culpability or the ability to blame someone wielding force because, like their victim, they too were simply in force's path. As Cha puts it, "Here is born the notion of a destiny under which executioners and their victims are similarly innocent: conquerors and conquered are brothers in the same misery."[5] The movement of force seems to bestow an equal blame, guilt, and responsibility upon all. In turn, this seems to demand an equality in our response to force through our ethical love. All are subjected to the objectifying power of force, and all have the possibility to respond with a decreative love that refuses to objectify the other person. In a sense, force evacuates the difference between guilt and innocence. Just as force renders all humans into objects, the decreative response to force renders all humans equally worthy of love. Both the negative (force, power) and positive (decreation, love) aspects of this movement are problematic if we remain in a mindset of crime and punishment or a concept of justice as a separation of the guilty from the innocent, and a subsequent punishment of the guilty. However, if one understands responsibility as prior to and independent of values like guilt and innocence, Weil's notion of force makes sense of intergenerational trauma and the cyclical nature of violence. It also recognizes the systemic prescription of roles and expectations that place one person in the role of oppressor and the other in the role of oppressed without even realizing that such a decision has been made by either party.

Moving beyond crime and punishment, and guilt or innocence, our ethical responsibility to the other person is prior to the other's actions. With this precondition, Weil's understanding of force is ethical. But if one remains in a mindset where ethics is applied after the fact as a way to distribute punishment, Weil's ethical structure dangerously and perhaps unjustly removes culpability from the oppressor. Furthermore, according to Lissa McCullough, Weil's decreative ethics expects love not just from oppressor to oppressed, but also from oppressed to oppressor (*RP*, 180). Again, this is a claim that is problematic within the mindset of ethical reward or punishment for actions but not if we think of responsibility to the other as prior to and separate from the divisive values of "guilt" or "innocence."

Roberto Esposito compares Weil's notion that force controls humans to Hannah Arendt's more traditional reading wherein humans control force. An ethics derived from the former is always going to be more complicated and problematic because human agency is so weak (*OP*, 44–45). Cha notes, "it is precisely here, in this position of radically uneven power born of one's exposure to force, that the key to resisting the notion of commanding power everywhere might be found and a relationship to the other that does not seek to harm him might emerge" (*DEB*, 42). The shared force between oppressor and oppressed contains within it the potential of grace, love, and recognition. This leads to what McCullough notes as a paradox in our ethical duty, namely, that we have no real will or power to control force, but we have an infinite amount of responsibility for the effects of force (*RP*, 162). Similarly, Cha argues that the necessary limit on our ability to give to the other does not in any way dispel our obligation to endlessly give (*DEB*, 16). Precisely because force's inclination is to run rampant, accruing more and more power, any ethical response to force through an attentive love of necessity will never be enough, but regardless, it must be done. While this may seem like a grim premise, I would argue that it is a form of optimistic realism in that it directly acknowledges a fact commonly ignored by ethical imperatives: most people will not follow those imperatives. If we do not recognize this going in, we may feel duped or that we have been treated unjustly. In Weil's estimation, we must go into our ethical actions acknowledging that there will be asymmetry and that our actions cannot ever fully overcome force. And yet, we must act.

Interestingly enough, while I use the term "optimistic realist," Scott Ritner is inclined to refer to Weil as a revolutionary pessimist, a stance that I see as similar to my own but approaching from the inverse disposition. Where we meet in the middle is Weil's combination of tireless ethical action and her realistic recognition that the problems she engages are unlikely to be solved.[6] I maintain my stance that Weil can be read as an optimistic realist for three main reasons: (1) The heart of Weil's work is love. While this love is admittedly achieved through both joy and affliction, her texts remain saturated in a possibility of divine love; (2) Following Frankl's notion of a "tragic optimism," I believe that Weil's philosophy gives us leave to say yes to life despite life's innumerable tragedies;[7] (3) Notwithstanding the evident calamity and angst of her brief life, I cannot help but read Weil as brimming with wonder and curiosity towards the world. I see this as one of her work's most compelling philosophical traits. To paraphrase the *Gita*, Weil's goal is the work and not the fruit of that work. As such, in acting, one has already achieved the highest goal, even if the fruits of her action leave much to be desired. This sense of achievement alongside the concept of a tragic optimism and divine universal love leaves me with a sense of optimism in Weil's work despite the realism in which she plays out the work of her thought. With that being said, I hold Ritner's account as equally valid and as arriving at the same destination in his estimation of Weil. The difference perhaps is simply which turn of phrase helps a person to hope beyond hope and love in the void.

Weil claims that it is *only* through my suffering that I am able to appreciate the suffering of the other: "Only he who has measured the dominion of force, and knows how not to respect it, is capable of love and justice" (*IF*, 212). Weil notes that suffering is like a koan (a Zen riddle meant to shift the mind out of its everyday perception into an enlightened state) offered to us by God: "God is the master who supplies this koan, plants it in the soul as something irreducible, a foreign body, impossible to digest, and constrains one to think of it" (*NB*, 484). It is interesting that Weil points to how *not* to respect force. She recognizes that at one point every human being carries the youthful pride of thinking that force will always be on her side. It is not until we lose force that we understand how unbalanced and unethically we have behaved. This goes both for the child in the school yard and the general at war. To be struck by force is to be confronted with the impossibility of certainty and an objectified, exhausted resignation. After this has happened, one cannot simply ignore the inevitable agony of existence.

In a sense, my own suffering or affliction, my reduction to the status of mere object by force, phenomenologically opens my ethical duties. We recognize the wisdom of Zosima in *The Brothers Karamazov*, who claims "Each of us is guilty before everyone for everyone, and I more than the others."[8] Our responsibility here is to halt the cycle of violence through the miraculous defiance of force. It is a miracle to have force on one's side and not to use it against another person. In this way, Weil agrees with Ivan Karamazov's position that the world is evil, but she disagrees that as long as one child suffers we can have no God (*NB*, 493). Only through Ivan's decreative action could God have access to ending the suffering of the innocent.[9] The only way for God to stop the suffering of that child would be through Ivan stopping the suffering of that child. He cannot blame God for that which he himself is choosing to do, or not to do.

The critique of Weilian force as an equalizer of all ethical culpability is dispelled when we consider Weil's ethical project as one of personal action and obligation in which the responsibility for the other is prior to and distinct from all other expectations and values. From this standpoint, the other person's blameworthiness or innocence is irrelevant in the face of her affliction. All that is relevant is one's own ethical action in response to that vulnerability. Weil believes that force can come into balance when we do not become wild with possessing it. She sees a type of balance in force but believes that such an equilibrium is lost in the war-riddled western culture in which she lived: "conceptions of limit, measure, equilibrium, which ought to determine the conduct of life are, in the West, restricted to a servile function in the vocabulary of technics" (*IF*, 195). In times of war, force runs rampant, and no human is left unscathed when she falls into the path of force.

Balancing Force

Force can be brought into balance by grace, but in order to understand this, one must understand the ontological position of Weil's God, who bestows that grace. In thinking through this concept, I use the pronoun "It" in reference to God because in my estimation this comes the closest to the type of God that Weil posits, and it avoids reinforcing patriarchal readings of a strictly masculine God despite that being the language used by Weil in her own writings. This is not an uncommon shift, as the use of the masculine pronoun to imply gender-neutrality has been increasingly abandoned in North American English. While "They" could be used as a singular pronoun for God in a way that is more personable for liturgical purposes, the use of "It" offers a freedom from anthropocentrism and allows for God to include a more impersonal Platonic notion of the Good. While "It" is not the most common pronoun for God, it has been utilized by thinkers from Pseudo-Dionysius to Sallie McFague.[10] Weil argues that "Nothing in the world is the centre of the world, that the centre of the world is outside the world" (*IC*, 174) and that "God, being outside the universe, is at the same time the center" (*WG*, 99). She explains this through the Christian difference and unity of God the Father and God the Son. For Weil, God's absence in the world is simultaneously Its presence. God the Father is totally absent from the earth and offers no reply in the face of affliction, but God the Son cries out the same "why" as all creation when he is crucified and seemingly forgotten on the cross: " 'Eli, Eli, lama sabachthani?' that is to say, 'My God, My God, why hast Thou forsaken Me?' "[11] For Weil, "The cry of the Christ and the silence of the Father together make the supreme harmony" (*IC*, 199). This harmony saturates the earth, infusing the world with beauty and suffering, affliction and joy,

God's presence, and God's absence, and with the paradoxical desire of creature (humans and God the son) longing for Creator (God the father, the Good).

For Weil, the choice to interrupt the cycle of violence is itself a miracle, as is any human choice that moves not with force and the accruement of more power but brings force into balance in and through grace. The miracle of disrupting the cycle of force is perhaps best described in the final verses of Homer's *Odyssey* when Athena, the goddess of war, is the only one able to end the bloodshed that has dominated the war itself and the warrior's long journey home. Odysseus slays the suitors who usurped his hearth during his long absence and finally finds himself home in Ithaca in the great rooted bed he shares with his wife Penelope. But in the final moments of the tale, he faces the wrath of the mourning suitors' families, and the cycle of war is about to continue its bloody reign:

> Odysseus and his gallant son charged straight at the front lines,
> Slashing away with swords, with two-edged spears and now
> They would have killed them all, cut them off from home
> If Athena, daughter of storming Zeus, had not cried out
> In a piercing voice that stopped all fighters cold,
> "Hold back, you men of Ithaca, back from brutal war!
> Break off – shed no more blood – make peace at once!"[12]

It is only the divine that can step in and offer the miracle of peace. Esposito notes the difference between Arendt, who suggests dialectically overcoming war through politics, and Weil, who argues for "overcoming" war by recognizing our own originary agonism (between human suffering via force and human love for the divine Good via grace) and attempting to bring this into balance through our political, economic, and social engagements in the world (*OP*, 5). Esposito also notes that for Arendt, war can be a method of asserting peace, whereas Weil sees war as a mechanism of reifying violence into the political system (*OP*, 43). It is only through the balance of grace that force can be recalibrated and justice can be achieved.

Weil refers to grace as the opposition to force. It is through grace that force can be ethically balanced. Recall that necessity serves to balance the natural and the supernatural, and as such, it is the meeting ground between force and grace. On the side of force is the energy inherent in "natural" matter that desires more and more power, and on the side of grace is the possibility of experiencing "supernatural" divine love in the beauty of the world, the love of the neighbor, religious practices, and friendship (*WG*, 83–142). Grace and force are not pure opposites, and they do not blot one another out. Grace and force cannot be synthesized, only balanced. Drawing on the Pythagoreans, Weil maintains opposing powers and allows them to remain open because it is only in acknowledging and allowing opposition that harmony is possible (*IC*, 162). Grace cannot prevent the misery of force, but it can help one to deal with that misery: "Grace can prevent this touch from corrupting him, but it cannot spare him from the wound" (*IF*, 214). Following Weil's explanation of force, I argue that grace is that *x* which flows from something beyond our material existence. Grace too is an unknown factor. Grace is a movement of truth and the Good, whereas force is the mechanical movement of matter towards power. In contradistinction, force is that *x* which exists in the material of the natural universe and annihilates human being's meaning and world. Grace connects to our highest potential towards love and truth, while force connects to our drive towards power.

This somewhat Manichean relation between force and grace resonates to a certain degree with the Cathars. Weil claims that "Catharism may be regarded as a Christian Pythagoreanism or Platonism; for in my eyes there is nothing above Plato."[13] And while Weil studied Cathar theology, her thought ultimately includes more differences than similarities to that of the Cathars.[14] The main difference between Weil and the Cathars around these two opposing principles of gravity or force and grace or God is that for Weil there is only one God who creates force through the divine self-abnegation or loving withdrawal, which allows the universe to exist. Esposito suggests that Weil's cosmology is influenced jointly by Isaac Luria and Hayym Vital's *tzimtzum*, Meister Eckhart's *Ent-Werden*, and myriad Platonic and Christian thinkers. Esposito notes that "rather than avoiding such contaminations, I find it more productive to emphasize the margin of originality and autonomy that contamination allows for in respect to these three traditions" (*OP*, 39). The Cathars, on the other hand, have a neo-Manichean belief that there are two originary Gods, or principles, one good and the other evil. While this could be read as similar to Weil's opposition between force and God or gravity and grace, it is metaphysically distinct in that it begins with only one: a God of love, who retracts Itself, creating the material force of gravity and building into that force the possible return to the one God. It is in force that the disrupting necessities of suffering, absurdity, and absence rattle the faculties to their very core, and it is only through grace and love that we are able to hold up both the faculty and its incapacity to provide absolute meaning.

It is worth reiterating here a point made in Chapter 1, namely that Weil's primary fidelity is to truth and not God. For Marie Meaney, Weil's readings are Christian, but Weil's ultimate tool of conversion is philosophy:

> [Pierre] Hadot compares philosophy to conversion. Philosophizing, especially with Socrates and Plato, presupposed a readiness to change, the wish to know oneself, and openness to truth. This is the attitude Weil desired from her readers, particularly when reading Greek tragedy.
>
> (*AL*, 23)

The spiritual tradition that she most closely aligned herself to was ultimately philosophy and the search for truth. However, Christianity, and Catholicism in particular, seems to be a close second for Weil. She offers a Christian apologetic reading of the Greeks, suggesting that the truth of Greek literature is aligned with the truth of Catholicism. Weil's sense of God more than anything reflects a vision of the Platonic Good realized through a Pythagorean harmony and linked to a reading of Catholic doctrine that has a strong implication of spiritual universalism.

As such, the presence of force and the absence of God on earth are both divine manifestations and not the product of the devil or human decision, such as Adam and Eve's fall from the garden. There is no original sin for Weil. Evil and hardship are a reality of embodiment necessarily brought about by God's loving decreation of Itself in order that the earth might exist. Force is a part of existence and is balanced by a grace that can manifest through human action.

The Impact of Necessity and Force on the Meaning of the Faculties

Upon reflection, it becomes evident that each faculty cannot exist as an absolute foundation of meaning in the world in which we live. The realities of absurdity, absence, and suffering in our lives compel us to recognize that while we may possess these faculties, it is impossible for them to offer meaning in an absolute and unadulterated way unless we live in a complete state of delusional imagination and self-deception. Indeed, even in setting out the faculties in the last chapter, their limitations had to be acknowledged by the philosophers I engaged: Hume distinguishes fact from the inevitable pull of values, Heidegger's Care is given meaning in our acknowledgment of the inevitability of death, and Descartes' limitless will is thrown into relief by the limited nature of understanding and thus human error. One can barely speak the faculty without referring to the limitation that in a sense, makes it what it is. As it is paraphrased by Gustave Thibon in *Gravity and Grace*,

> The irreducible character of *suffering* which makes it impossible for us not to have a horror of it at the moment when we are undergoing it, is destined to bring the *will* to a standstill, just as *absurdity* brings the *intelligence* to a standstill, and *absence love*, so that man, having come to the end of his human faculties, may stretch out his arms, stop, look up and wait.[15]

We are forced to face both the faculty and the opposing necessity of existence. Both equally exist in a blatant opposition that refuses a total domination by one side. Recognizing this can initially be world-rending and painful but it ultimately allows for a decreative shift away from the *ersatz* expressions of these faculties and into productive ethical action. In considering Weil's fascination with both passivity and work, Simone Kotva notes that for Weil,

> Whatever I do in the world – however cleverly I seem to grasp or lay my hands on it – I remain subject to it because the world, like my body, is never wholly within my control. Of course, in my imagination I may be able to "turn the world upside down," as she writes, while meeting no resistance, but this idea is a fantasy.[16]

The work of force and necessity guide us to realize that we have some control but that control admits to limits and as such, our faculties cannot impose an absolute meaning or truth upon the world.

The Necessity of Absurdity (Contra Knowing)

Each faculty finds meaning through the engagement with and employment of patterns of force, force that, however, we often forget, exceeds human knowledge and control, something we are reminded of when we run up against necessity. While a faculty is upheld by force, it feels invincible. In the case of the faculty of knowing, Dr. Faustus' acquisition of unlimited worldly knowledge is the perfect example in that it supposes the possibility of such knowledge through the arrogance of force and simultaneously sees the impossibility of this absolute knowledge in that it requires Faustus to bargain with the Devil.[17] The opposite of knowing can be found in absurdity, which is not a strictly negative possibility or a property of force, but rather a recognition of necessity's limitation upon force. Paired with knowing, absurdity can act as a potential avenue to divine ethical Good if one is attentive in thinking between these two contradictions. On the other hand, if one allows oneself to become totally overwhelmed by the absurdity of existence, one is drawn into the sway of force and becomes a catalyst for the unethical. Dr. Faustus is as in danger of being overwhelmed by the hunger for absolute knowledge as he is of being overwhelmed and even made suicidal by the absurdity of existence and the meaninglessness of that knowledge and power he has attained when it comes to creating meaning in the world.

The faculty of knowing, as I have established it in the previous chapter, forms a firm foundation dismissing all that is beyond rational logic and empirically verifiable facts. In purely hypothetical or strictly rational notions, the faculty of knowing can totalize and understand all things. But it is from a firmly in-worlded position that Albert Camus claims, "At any street corner the feeling of absurdity can strike any man in the face."[18] Absurdity is an inevitable part of our existence that shatters the absolute foundation of knowing, but it is also a necessity that allows us to recognize our own limitations. Weil explains:

> Our life is nothing but impossibility, absurdity. Each thing that we desire is in contradiction with the conditions or the consequences attaching to that thing; each assertion that we make implies the contrary assertion; all our feelings are mixed up with their opposites. The reason is that we are made up of contradiction, since we are creatures, and at the same time God, and at the same time infinitely other than God.
>
> (*NB*, 411)

For Weil, absurdity is a certainty of embodied existence. This absurdity lies in the gap between what we "know to be true" and "how things are in the world." Religious mysteries, for example, are all absurd and non-sensible and as such, they should be denied at an intellectual level but allowed to exist at the level of emotional spiritual well-being, as something to allow for the full person to flourish, akin to great music or paintings, which Weil argues give nourishment to the soul (or the spiritual emotional aspects of a human) and relatively little to the intellect proper.

Ultimately, when it comes to the question of life's meaning, we have moved beyond empirical or rational knowing. The question of meaning is impossible to answer rationally or logically. Prince Hamlet may have found himself 30 years old and still studying at Elsinore, but he could not find the answer to that most basic of questions: "To be or not to be?" Knowing in the positivist sense can offer us the particular (and sometimes even general) "how" but never the universal "why." Weil claims, "The principal effect of affliction is to force the soul to cry out 'why' as did the Christ himself, and to repeat this cry unceasingly, except when exhaustion interrupts it. There is no reply" (*IC*, 198). Absurdity is the recognition that we cannot know the ultimate answer to this most fundamental question. Thus, on an ontological level, knowing finds itself inadequate as a foundation by its own definition as a meaning that is logical, rational, and empirical. The faculty knows particular ontic things, but it knows nothing of life's ultimate meaning because that exists beyond the realm of empirical knowledge.

Furthermore, as the deconstruction of Jacques Derrida has shown us, truth can always be inverted, told from a different perspective, and reimagined in myriad ways.[19] Scientific "truth" is the best theory we have and not an *Absolute Truth*, as logical positivists might suggest. Indeed, while Hume and the Vienna Circle carefully distinguish fact from value, Catherine Elgin notes that many facts are formed and based upon values.[20] This is not an ethical license to ignore science, as it is not a glib suggestion that science is truly the best theory we have; rather, it is an epistemic claim about the nature of scientific truths and the impossibility of objective Truth.[21] Unfortunately, the inability to see the nuance between an epistemic and an ethical claim leads postmodernism to be considered a post-truth movement without morality rather than a recognition of the delicate and fluid nature of truth. Within the current social political climate, it is perhaps not correct to say that the nuance is lost, but rather, it seems that it is politically or socially expedient or lucrative for some academics and public intellectuals to ignore this nuance in favor of a simplified version of postmodernism. But be it by genuine misunderstanding or in pursuit of a political agenda, it seems that postmodernism's lessons about truth are commonly overlooked, which is unfortunate when we consider the importance of recognizing the necessity of absurdity in our day-to-day lives. Knowing is paralyzed by absurdity when we recognize knowing's inability to provide meaning in contrast to the fundamental human desire to find meaning and create worlds. In this way, the faculty of knowing is interrupted by the necessity of absurdity.

The Necessity of Absence (Contra Loving)

The faculty of loving is a foundation that fills our lives with responsibility, care, worry, and joy. It is a foundation that places its meaning in other beings, objects, and places. Life is made meaningful by loving and ordering those things. One directs love towards particular bodies in the world and one's life is given meaning because of those relations to bodies. The faculty of love, however, meets its match in the necessity of absence. Love becomes paralyzed when it runs up against absence. Absence reveals itself in myriad ways through which the particular object loved is lost, but perhaps the most accessible to conceptualize is death. As in the case of C.S. Lewis and his wife, Joy, whom he writes about in his beautiful essay *A Grief Observed*, one may plan to spend a lifetime with a certain partner, but the untimely death of that person cares not for the arrangement and there is no amount of loving that can change this reality.[22] Meaning that is placed on a particular being is radically altered with the death of that being. All that exists in time and space is impermanent and thus there is an incongruity in directing all of our loving towards that which will inevitably be gone.

When that which we love is gone or even if we simply conceptualize the inevitability of its disappearance, we experience the void: "To love truth means to endure the void and, as a result, to accept death. Truth is on the side of death" (*NB*, 160). Love does not end upon the death of the beloved, but the direction of love is forced to shift. Loving can no longer be directed to the presence of a physical body and often dissipates or is reimagined. McCullough notes that our acceptance of death leads to a shift from desire directed towards "life" to desire directed towards the "good" (*RP*, 69). When love reaches this limit, we are forced to enter the void. Our love, directed at a particular body (a beloved family heirloom jewel, a lover, a favorite nook of a tree, a child), is paralyzed by the absence of that entity either in its annihilation as a body, or in our inability to engage with it. The beloved lost to indifference (a breakup, a divorce, a parting of ways) and the beloved lost to death similarly rend the lover in that the object of their love is no longer accessible.

In Weil's kenotic theology, the void is the necessity upon which the world is created: by retracting Itself and creating a lack, God makes space and time in which the universe can exist. When we enter the void, losing the firm foundation of the faculty of love, Weil is adamant that we should not imagine any consolations or allow the imagination to soften the desperation and pain of experiencing this dark night (*NB*, 148). It must be endured without hope of consolation. It is important to note that while these necessities of existence may seem "negative" in that they destroy our foundations of meaning, this is actually an incredibly important step in our quest for truth, the possibility of decreation, the recognition of our self as part of a larger sense of being, and ultimately, for Weil, the possibility of God. But this ultimate consolation of God is only available if we do not dare to hope for it. The arresting of the faculties by necessity has the potential to act as the dark night of the soul through which we may come to a new depth of meaning and wonder in our existence. There is a necessary atheistic openness in this step, in which we must open to the possibility of absolute nothingness in order to have no preconceived notion of the divine Good that we may find there. As such, the faculty of loving is extinguished by absence. The particular beloved is absent, but so too is the consolation of a divine comfort. One loses the ground of this faculty and must stand completely alone in the void in order to find her way back.

The Necessity of Suffering (Contra Willing)

Weil argues that human will is not infinite and that it must adhere to the laws and limits of physical necessity. Suffering reveals that the human will is bound to the conditions and limits of reality. The suffering of factory workers and of slaves, the prevalence of wars and oppression, the affliction of the innocent all show us the limitations of our willing in the world. A person may go to college for a particular degree with the idea of attaining a job, but a financial crash and the subsequent economic depression that makes working in that field impossible does not heed her willing. Perhaps a loss of other family funds forces her to work in a menial job for minimum wage in order to support her family in the immediate present, and thus the future possibilities of working in her desired career grow fainter and fainter. The crashing in of suffering upon one's life reveals the limitations of willing. Our contemporary neoliberal capitalist society constantly peddles this foundation of willing when it promotes the "American Dream" or the idea that with courage and willpower, one can "do anything." But the reality of financial disparities between poverty and prosperity, even in wealthy democratic nations, shows us the limits of the will in the wake of suffering. This is not to say that the limitations imposed by neoliberal capitalism are inevitable in the same way that death is inevitable. Rather, neoliberal capitalism is a catalyst for force akin to war in its ability to ravage all in its path and allow force to operate unbridled. As such, it can be the harbinger of suffering, but it is not in any way predetermined. The necessity of suffering as a limitation upon our willing is exemplified in Robert Burns' lines:

> But Mousie, thou art no thy-lane,
> In proving foresight may be vain:
> The best laid schemes o' Mice an' Men
> Gang aft agley,
> An' lea'e us nought but grief an' pain,
> For promis'd joy![23]

In this short verse, Burns notes that human willing, like that of all creatures, is not enough to ensure the joy that we seek, and in fact, the results of our willing may be completely superfluous or counterintuitive to reality. John Steinbeck brings this sentiment into the twentieth-century social political ideals of America in his novel *Of Mice and Men.* Willing as a mode of meaning making is rendered impossible by the worldly necessity of suffering, and the best-laid plans of mice and men so often fall short.

When the faculties are thus paralyzed by the necessities of force, the human comes to a fork in the road. If she has cultivated a special attention to them, she can breach the horizons of everyday rationality and enter into the realm where the impossible and the possible are not, strictly speaking, mutually exclusive but can exist together. However, if she has no capacity for attention, she will fall into a pitiless despair at the enormous loss of meaning:

> The void that one grasps between the pincers of contradiction is indubitably the one which lies above, for the more one sharpens the natural faculties of

> intelligence, will and love the better one grasps it. The void which lies below is the one into which one falls by allowing the natural faculties to become atrophied.
>
> (*NB*, 412)

Through cultivating attention towards and within the faculties, we make the most of these inevitabilities of force and find the grace that does not spare us from the whips and scorn of existence but offers a way forward. Without fine-tuning our faculties, we fall into nihilism and despair. The faculties are paralyzed and rendered meaningless, leaving the individual to either embrace nihilism or alternately to step into absolute unfettered possibility.

Weil answers the destruction of meaning with the possibility of *metaxu*, or a fleeting connection between human and divine via love. According to Iris Murdoch, the use of *love* is what is so appealing in Weil and what she believes gives Weil's philosophy a true potential morality in the face of destruction and loss.[24] As Weil argues, "You could not have wished to be born at a better time than this when everything is lost" (*FLN*, 47). With an ethical inversion of this principle, Naomi Klein observes in *The Shock Doctrine* that radical loss is always also radical opportunity.[25] For Klein, this is a matter of capitalist companies taking advantage of natural disasters in order to buy up property and change the systems in place. Disaster and loss are always also opportunities, and these opportunities can be used in radically unethical ways (such as the corporate greed that Klein describes) or radically ethical ways (such as Weil's decreation). In the next chapter, I will turn to what *attention* precisely means for Weil in order to understand how it allows for the faculties and their impossibility to exist at once in an open mediation.

Notes

1 Nicolas Bommarito is currently working on an edited collection entitled *Dharma and Detachment*, which takes the passages from Weil's notebooks inspired by Eastern philosophy and brings them into an edition similar to *Gravity and Grace*.

2 Eknath Easwaran, trans., *Bhagavad Gita* (Boston: Shambhala, 2004), 3:27–28.

3 Florian Cajori, *A History of Mathematical Notations* (New York: Dover, 2011), 191.

4 For more, see Thomas M. Kostigen, *Hacking Planet Earth: How Geoengineering Can Help Us Reimagine the Future* (New York: Tarcher, 2020); Oliver Morton, *The Planet Remade: How Geoengineering Could Change the World* (Princeton: Princeton University Press, 2017). Against this geoengineering solution, see Michael Huesemann and Joyce Huesemann, *Techno-Fix: Why Technology Won't Save Us or the Environment* (Guelph: New Society Publishers, 2011).

5 *DEB*, 42. Quoting Weil, "Iliad," 31.

6 See Scott Ritner, "Simone Weil's Heterodox Marxism: Revolutionary Pessimism and the Politics of Resistance," in Sophie Bourgault and Julie Daigle, eds., *Simone Weil, Beyond Ideology?* (New York: Palgrave Macmillan, 2020), 185–206. Ritner is currently working on a full manuscript on this topic entitled *Revolutionary Pessimism: Simone Weil on Politics*.

7 Viktor E. Frankl, *Man's Search for Meaning* (Boston: Beacon Press, 2006), 161–179.
8 Fyodor Dostoevsky, *The Brothers Karamazov*, trans. Richard Pevear and Larissa Volokhonsky (New York: FSG Books, 2002).
9 See *RP*, 107 and *AL*, 162–163.
10 Pseudo-Dionysius, "The Divine Names," in *Pseudo-Dionysius: The Complete Words*, trans. Colm Luibheid (New York: Paulist, 1987), 54–56; Sallie McFague, *The Body of God: An Ecological Theology* (Minneapolis: Fortress, 1993), 147.
11 Wayne A. Brindle, et al., eds., *The Holy Bible. New King James Version* (Nashville: Thomas Nelson, 1988), Matthew 27:45–47.
12 Homer, *The Odyssey*, trans. Robert Fagles, in *The Norton Anthology of Western Literature*, eighth edition, vol. 1, ed. Peter Simon (New York: W.W. Norton, 2006), Book XXIV lines 579–585, p. 495.
13 Simone Weil, *Seventy Letters*, trans. Richard Rees (Oregon: Wipf and Stock, 1969), 131.
14 See Stephen Plant, *Simone Weil: A Brief Introduction* (London: Harper Collins, 2007), 87–102.
15 *GG*, 166. My emphasis. I cite this as written by Thibon rather than Weil because I have been unable to find the corresponding citation in Weil's *Cahiers* from which Thibon took the content that now makes up *Gravity and Grace*. As such, it would appear that this is not Weil's writing but Thibon's own interpretation of it, as he did add in and considerably rearrange Weil's notebooks in order to create *Gravity and Grace*.
16 Simone Kotva, *Effort and Grace: On the Spiritual Exercise of Philosophy* (London: Bloomsbury, 2020), 135.
17 Johann Wolfgang von Goethe, *Faust: A Tragedy in Two Parts*, trans. Thomas Wayne (New York: Algora Publishing, 2016).
18 Albert Camus, *The Myth of Sisyphus* (London: Penguin Books, 2005), 9.
19 For more on Derridean deconstruction, see Jacques Derrida, "Différance," in *Margins of Philosophy*, trans. Alan Bass (Chicago: University of Chicago Press, 1982), 3–27.
20 Catherine Z. Elgin, "The Relativity of Fact and the Objectivity of Values," in *Between the Absolute and the Arbitrary* (Ithaca: Cornell University Press, 1997).
21 For more on the problematic nature of objectivity, see Lisa M. Heldke and Stephen H. Kellert, "Objectivity as Responsibility," *Metaphilosophy* 26.4 (October 1995), 360–378; Catherine Z. Elgin, "Making Up People and Things," in *Between the Absolute and the Arbitrary* (Ithaca: Cornell University Press, 1997).
22 C.S. Lewis, *A Grief Observed* (Harper One: New York, 2015).
23 Robert Burns, "To a Mouse," in *The Complete Poems and Songs of Robert Burns* (Glasgow: Gresham Publishing Company, 2015), 83.
24 See Iris Murdoch, "On God and the Good," in *The Sovereignty of Good* (London: Routledge, 2001), 65.
25 Naomi Klein, *The Shock Doctrine: The Rise of Disaster Capitalism* (Toronto: Vintage Books, 2007).

4 Attention and Mediation

Attention: The Balance of Faculty and Force

Weil likens attention to a form of prayer in that it is an openness to the surrounding natural world that can lead one to a transcendent supernatural sense of meaning, the Good, and the possibility of the ethical (*WG*, 57). Sharon Cameron notes that while attention may lead us to the supernatural and is a form of prayer, in itself it is a natural act: "For while virtue is a supernatural thing, attention for Weil is a natural one."[1] Attention is a natural act that can lead one to *metaxu*, which links natural and supernatural. Attention is not a muscular effort and is not attached to or driven by the faculty of willing beyond the will's negation or the will-not-to-will. Instead, Weil refers to it as a different kind of effort: "Attention is an effort, the greatest of all efforts, perhaps, but it is a negative effort" (*WG*, 61). Rather than asserting our will, attention requires an openness:

> Attention consists of suspending our thought, leaving it detached, empty, and ready to be penetrated by the object; it means holding in our minds, within reach of this thought, but on a lower level and not in contact with it, the diverse knowledge we have acquired which we are forced to make use of.
>
> (*WG*, 62)

This openness or mindfulness of attention thus allows for other things to *be*, to open, to express. Similar to practices of meditation or prayer, this negative effort is an attempt to stop the mind from actively dominating and appropriating each thing that it encounters. As such, attention has an ethical dimension as well: rather than forcing our own totalizing and often predetermined understanding onto other things and beings, attention allows for the meaning of the world to be co-created with others. Simone Kotva argues that for Weil, "A person is of necessity neither disembodied mind nor mindless matter but an embodied life; she is active and passive at the same time but not in the same way; and meditation – practised attentively – can initiate the philosopher into a realization of this state."[2] Cultivating attention is a way to enter into the wholeness of existence, to open up to the combination of action and passivity that marks the human experience.

DOI: 10.4324/9781003449621-6

Each time we bring open attention to something, we refine our aptitude to be attentive. This ability is available to all people through dedication and discipline and yet is rarely put into practice, according to Weil. By bringing our attention to the three faculties, we recognize the impossibility of them, and thus when we are inevitably struck in the face with absurdity, absence, and suffering we have the ability to recognize both the faculty and its impossibility at the same moment. Through attention we are able to avoid falling into despair and losing all meaning, because we can recognize the presence of the realities of existence *and* our faculty's ability to adapt to these realities through the mediation that this chapter will unpack. In this way, we can enter into the ethical action that had seemed impossible when we thought of only the *ersatz* faculty or the necessity that paralyzes our meaning:

> We have to accomplish the possible in order to be able to seize upon the impossible. The right way of exercising, in accordance with duty, the natural faculties of the will, the intelligence and love corresponds exactly, in the case of spiritual realities, to what the movement of the body is in relation to the perception of sensible objects.
>
> (*NB*, 417)

Only attention can offer us a window into our duty and the correct ethical actions that we ought to take in life. Cha argues that our ability to pay attention is not simply this openness but is, more importantly, a form of moral aptitude (*DEB*, 18). In the same way that an inattentive person bumps into things, knocks things down, and mishears conversations in the physical realm, the inattentive person misreads her duty and the use of her faculties in the ethical realm.

Different from the Aristotelean distinction between a lack of self-control (*akrasia*) and the ability to assert self-control (*enkrateia*), attention does not make an ethical judgment, it simply steps back, allowing the person, being, or thing where attention is focused to become present in ways that are not overwritten and obscured by desires and preconceived understanding.[3] This takes an immense amount of restraint (or rather self-abnegation) on the part of the attentive subject, but it is not with any preconceived notion of a good in mind, because this would, of course, be yet another mode of the willful mind controlling and totalizing the experience.[4] For Aristotle, self-control involves the capacity to master motivations that result in behavior contrary to one's better judgment. A. Rebecca Rozelle-Stone and Lucian Stone set out how akrasia "turns out to be surprisingly relevant to our understanding of Weilienne love" in that it may be radically reimagined: "rather than being a sign of moral degradation, it is a condition of humility, of (literal/physiological) openness to the world, that is, for Weil an ethical receptivity."[5] The Rozelle-Stones suggest that such a rethinking of *akrasia* would align Weil's ethics with this notion of Aristotelean lack of self-control as a positive ethical possibility rather than a negative ethical privation. Attention, then, cultivates a release of the illusion of control rather than cultivating a willful control. Weilian attention might be seen as an assertion of the totalizing self to achieve a higher ethical good, but ultimately, it does not step in and guess at or judge what the ethical answers may be from a position of mastery over self or over the other being or thing. It is, in a sense, a form of attentive openness to the other through self-abnegation.

Through the correct attention to the three faculties and the subsequent three modes of existential necessity, we are overwhelmed not by the evil of the world as we would be in a Cathar or Manichean metaphysics but by the question *why:* "it is only after we have exhausted all the natural faculties in us (will, intelligence, natural tendency towards loving) in the effort to produce good, and have recognized that we are incapable of good, that we fall prostrate before God" (*FLN*, 354). The question "Why?" is followed by God's silence: "when we cry out for an answer and it is not granted us – it is then that we touch the silence of God" (*NB*, 627). Here, we are finally dumbfounded with the impossibility of our existence and thus, perhaps, of God: "Impossibility – that is, radical impossibility clearly perceived, absurdity – is the gate leading to the supernatural. All we can do is to knock on it. It is another who opens" (*NB*, 412–413). In this silence of God, we have the ability to recognize God's absolute absence on earth, and yet, "The apparent absence of God in this world is the actual reality of God" (*NB*, 424). For Weil, the universe is made through God's love and Its decreative removal of Itself so that we may exist. At the heart of God's (the Father or the omnipotent) earthly absence and the emptiness of existence, we find the love and grace of God (the Son or the suffering fellow being) (*FLN*, 338).

The Balance of *Metaxu*

The ability to experience the supernatural in the natural world occurs through Weil's concept of *metaxu* (μεταξύ) or the notion of the *middle-ground* or *in-between* taken from the discussion of Diotima's wisdom in Plato's *Symposium*.[6] This concept emerges in *Symposium* with Socrates' suggestion that "there is no desire if there is no lack"[7] and from the conclusion that "what Love lacks and does not have is beauty" because love is the desire for beauty.[8] Love is described by Diotima as "something in between (μεταξύ) mortal and immortal" and "He [Love] is a great spirit … All spirits are intermediate between god and mortal."[9] Marie Meaney posits that for Weil, as for Plato in his famous ladder of love, beauty "is the only transcendental accessible to the senses, it is the only one which, once inside the cave of Plato's myth, can tempt the prisoners to break their chains and eventually discover the good outside" (*AL*, 63). By transcendental, Meaney is referring to something which "has its origin beyond this world and manifests God" (*AL*, 63). This transcendental manifestation of God in the world through attention to beauty or the glimpse of the good through ethical action is also described as the "supernatural" in Weil.

Metaxu introduces the idea that "Every separation represents a bond"; it is a way to momentarily and partially overcome the chasm between the supernatural Good or the creator and the natural earthly human creature, allowing them to recognize their interconnection (*NB*, 497). Eric Springsted posits *metaxu* in Weil as the links between the quotidian and the divine.[10] Springsted describes *metaxu* as "things, states, or activities within this world which are subject to the correlates of good and evil … they are things which can in a limited way be used to bring the soul into contact with God."[11] These objects are not objectively *metaxu* but become so in a combination of the object as a harbinger of the good *and* one's attention to said object. On the one hand, the Lord's Prayer was written to be such a link and has been used expressly for this purpose, which gives it an extraordinary potential to act as a connection. On the other hand, if my own attention does not meet the prayer, it will never become a connection for me. But if my attention is engrossed not in Scripture but in, for example, George Herbert's poem *Love III*, then this poem may act as *metaxu*. Indeed, Herbert's poem brought Weil her own encounter with the divine (*WG*, 27). Peter Winch notes that Springsted's reading of *metaxu* pushes the concept beyond mere metaphorical connection to the divine and offers a possibility of metaphysical transcendence.[12]

In utilizing our attention to hold both faculty and necessity in balance, we have already begun the process of decreation and worldly action. Giving attention to the faculties and their impossibilities cultivates a deep shift in how we conceptualize our faculties. Through attention, each faculty undergoes a transformation that is a part of the process of decreation. The process of decreating the self is an ongoing intertwining of natural and supernatural. This movement does not leave behind the natural but acts as an open mediation that reveals the interconnection of the two. The first step of this movement is the aforementioned attention to our faculties and their worldly impossibility. Attention begins to develop the evolution that constitutes decreation.

Wisdom as an Open Mediation (Between Knowing and Absurdity)

In a sense similar to the empiricist acquisition of knowledge, for Weil, knowledge is a matter of attention. But completely unlike the fact and value division seen in the Vienna Circle, Weil sees the exercise of acquiring knowledge or fact as a pathway to the value system of God through ethical actions. Joseph Cosgrove explains that for Weil, philosophy has the potential to contribute to positivist knowledge in a meaningful way by engaging the "merely utilitarian or technological conception of science" with a "genuine conception of truth and goodness."[13] Only by tempering a positivist knowledge with critical philosophy can one overcome propositional truths to attain a science of truth as Goodness. Weil sets out an attention-based doxological (meaning the praise of the divine) form of knowledge acquisition. For Weil, it is the "faculty of attention" that "forms the real object and almost sole interest of studies" and which forms the foundation of prayer (*WG*, 57). One pays attention by "suspending our thought, leaving it detached, empty, and ready to be penetrated by the object; it means holding in our minds, within reach of this thought, but on a lower level and not in contact with it, the diverse knowledge we have acquired which we are forced to make use of" (*WG*, 62). Additionally, Weil notes that we must dwell upon our epistemic failures, tracing each fault back to its origin in our mind. And finally, intelligence and epistemic pursuit must be fueled not by willpower but by joy: "The intelligence only bears fruit in joy" (*WG*, 61).

This joyful pursuit of knowledge through attention and an acknowledgment of one's own failures prepares the student for loving God through attentive prayer with the ultimate goal of ethical action. School studies done in this way "can one day make us better able to give someone in affliction exactly the help required to save him, at the supreme moment of his need" (*WG*, 65). Thus, Weil sees the exercise of acquiring knowledge or fact as a pathway to the value system of God through ethical actions. Here Weil layers the work of the philosopher with the work of the mystic. Kotva notes that for Weil:

> A phenomenon like attention – the mystic's principal tool, but also the method of the philosopher – seems to contradict all notions of passivity, since in order to manifest attention, one must be alert and vigilant, and yet it is only when a person is "tired and [thinks] … almost dreamily, unconsciously," that what they are seeking for finally becomes apparent.[14]

Weil posits knowledge in the Humean sense as important, but that knowledge is operating at a separate and lower level than the *wisdom* of open attention, which is "the greatest of all efforts perhaps, but it is a negative effort" (*WG*, 61).

For Weil, I have suggested, the *ersatz* faculty of knowing consists of the logical truths of empiricism but also the know-how of common sense. By definition, it is limited to the natural world in its scope and understanding. But according to Weil, the "natural" world is not the whole story when it comes to human existence. Diogenes Allen argues that in her spiritual experience of Herbert's poem, Weil is able to experience something beyond mere intellect.[15] *Wisdom* carries with it the connotations of both the natural and the supernatural. It is the ability to see the logical empirical truth as one part of a larger whole. It does not deny or reject knowledge, but rather understands the order of those truths. Knowing can only cross over into wisdom when it recognizes the necessity of absurdity, or as Vance Morgan notes:

> An important lesson to be learned from scientific enquiry is that the world is not at our disposal: rather, we are in many ways at the mercy of the necessity of the world. Recognition of this fact is, for Weil, one of the most important preliminary steps to humility and wisdom.[16]

The shift into wisdom is an achievement of holding open absurdity alongside knowing.

Weil sees the zenith of wisdom in the Pythagoreans. While Weil takes necessity to be the object of math and science (she sees math as *the* science of nature) (*IC*, 181), she also notes that it is not inspired by the natural, but by the supernatural: "in science, as in art, all true novelty is the work of genius; and true genius, unlike talent, is supernatural" (*IC*, 171). To be a genius, for Weil, is to pay careful attention to the natural world with an openness to the supernatural possibility of goodness. For Weil, the problem with knowing common to modern positivist science is that it is only interested in the natural: "One does double harm to mathematics when one regards it only as a rational and abstract speculation. It is that, but it is also the very science of nature, a science totally concrete, and it is also a mysticism, those three together and inseparably" (*IC*, 191). The reality of the world is such that we are forced to rethink the totalizing grasp of our knowing, and as such, we begin to allow in that which exceeds the limits of our own knowledge and to make space for alternate ways of knowing. This is the inception of what I have called wisdom and the decreative open mediation between knowing and absurdity. Morgan explains that "[a]lthough the world presents us inexorably with obstacles to overcome, confronting us with necessity that burdens our very existence, the beauty of this very same world draws us toward all that is of value and goodness."[17] The only way that we can arrive at wisdom is moving through knowing and absurdity. The hollowness of a promise of total logical control and meaning from a world of indifference must be felt in order to be struck by the beauty and goodness also within that world.

Weil places mathematics as a science in this Pythagorean harmony, which notes the distinction between the sciences and the humanities but insists that thinkers must always seek to find *metaxu* linking our knowledge even in diverse fields. Cosgrove argues that this form of "mystical science" that seeks bridges in our ways of knowing is incredibly important as a mode of connecting the sciences and the humanities.[18] The harmony born of aporia, when carefully studied through sciences and mathematics, brings about the recognition of the beauty of the world saturated in the love of God: "The beauty of the world is God's own beauty, as the beauty of the body of a human being is the beauty which belongs to that being. But wisdom, righteousness and the rest cannot appear to us in the world but only in a human being who will be God" (*IC*, 150). Thus, there is not an absolute division between nature and God but rather a possibility of *metaxu* held open by the divine earthiness of other human beings. Wisdom, then, is the movement from the empirical emphasis merely upon the concepts of gravity in one's knowing to include the concepts of grace through wisdom.

Loving God as an Open Mediation (Between Loving and Absence)

Weil's mediated higher notion of love is ethically social-justice-oriented like bell hooks' and layered over the ontological inescapability of concerned involvement found in Martin Heidegger. Weil, though, has two layers of love at work in her thinking: the love of the particular and the love of God. The fallible *ersatz* faculty of love, as I am setting it out in Weil, denotes the first type of love, namely love for particular embodied, earthly beings. The second type of love, as a love of God, has the possibility to act as an open mediation towards ethical action. The faculty of love, as the love of particulars, is explained by Weil as the love of necessity, or loving the reality of our earthly existence (*IC*, 186). For Heidegger, this reality of earthly concerned involvement is an existential condition, but to love this condition itself better conceptualizes what Weil's love of God might entail. To love necessity, for Weil, means to pay attention to the surrounding world and to notice necessity's participation in constraint, justice, beauty, and faith. In paying attention to the world, in noticing it, we see our connection to it, and it becomes something that we can love (*IC*, 188). In this way, loving expands ethics.

The first kind of love (for particulars) moves into the second kind of love (for the universal Good) and for Weil they are not different in kind but rather in degree: "Not only does the love of God have attention for its substance; the love of our neighbor, which we know to be the same love, is made of this same substance" (*WG*, 64). As such, Weil refers to both forms of love as a love of God, but it should be noted that this first form "is the love of God, whether or not the name of God be present in the mind" (*WG*, 175). She explains that this first form of love happens through loving earthly particulars, and she argues that there are four possible earthly vehicles for this love: religious ceremony, beauty of the world, our neighbor, and friendship (*WG*, 83). One must love the earthly elements of existence before one can have any possibility of experiencing the love of God (*WG*, 84). Similar to the social justice imbedded in bell hooks' ethics of love, Weil closely links love of the neighbor with justice. To truly experience another person is an "absolute identification of justice and love" for Weil and this justice comes about through finding balance: "The supernatural virtue of justice consists in behaving exactly as though there were equality when one is the stronger in an unequal relationship" (*WG*, 85–87).

In Weil, loving begins as loving a particular natural material object or being in the world. Then, following the encounter with absence, love has the potential to shift into loving God. There is not a difference of kind in these two loves, but of degree or direction: "love is a direction and not a state of the soul. If one is unaware of this, one falls into the onslaught of affliction" (*WG*, 81). The absence of a love lost can lead to terrible suffering if one is unable to redirect that love from the particular thing (now inaccessible) to the universal divine Good. Only when love of the natural world is directed freely and without hope is there the possibility of the supernatural love of God. This is why Laurie Gagne posits that "With their selfless efforts to preserve the natural world, Weil would say, even many an avowedly secular activist is in fact loving God unawares."[19] As such, the concept of God is particularly aligned to doing ethical Good in the world for Weil. This notion is taken up in an atheistic vein by Iris Murdoch when she suggests that attention is a form of prayer without God: "Prayer is properly not petition, but simply an attention to God which is a form of love. With it goes the idea of grace, of a supernatural assistance to human endeavour which overcomes empirical limitations of personality."[20] For Murdoch, moral philosophy must maintain the concept of "God" without the religious baggage: "God was (or is) a single perfect transcendent non-representable and necessarily real object of attention; and I shall go on to suggest that moral philosophy should attempt to retain a central concept which has all these characteristics."[21] God, or for Murdoch, the concepts under the category of God, offers a higher moral possibility in the notion of love.

So how can we differentiate between the love of the earth/humans, as the first love of the individual person or thing that can be destroyed by absence, and the love of God as this second universal moral love? If someone can move into the love of God without recognizing it as God, there seems to be a lack of clarity about this shift. The love of God does not overtake the love of the world, but rather enhances it:

> those whose love of God has caused the disappearance of the pure loves belonging to our life here below are no true friends of God … Our neighbor, our friends, religious ceremonies, and the beauty of the world do not fall to the level of unrealities after the soul has had direct contact with God.
>
> (*WG*, 142)

Thus, the first love of earthly objects is enriched by the second love of God but there is not a hard and fast ontological difference between the two. The love of God allows for a depth in our love of the world but does not alter that love in kind, only in degree. Lissa McCullough notes:

> Neither the one (love directed to creatures) nor the other (love directed to God) may be viewed as the uniquely decisive end apart from the other. Each movement of love is thoroughly imbued with the other. We must not prefer one movement over the other; rather we must unite them in the way we love, so that our love of creatures is purified by our love of God, and our love of God is held to its earthly criterion in the love of creatures.
>
> (*RP*, 201)

Roberto Esposito reads Weilian love as the "non-force more forceful than force" and as the path towards harmonious justice (*OP*, 68). Both loves are beautiful. The love of the world is beautiful in that it is a manifestation of pure necessity, which is God's absolute absence. This reveals to us the perfect love of God that withdrew Itself to create the world. Thus, the universal love of God is at the core of pure necessity and of the beauty of the world, which manifests as the love of the particular. According to Joke Hermsen, in classic negative theology, God is present on earth but any positive statement about God is inadequate due to our human limitations. On the other hand, Weil states that God abandons the world and relinquishes being everything. Thus, God creates the world as neither everything nor as nothing.[22] Hermsen argues that affirmation of the world is necessary for Weil's conception of God as beyond the world: "this brings us to a recurrent paradox in Weil's thinking: only through affirming the world can we trace what is beyond that world."[23] Consider how love changes towards a lover upon her death. The love does not vanish, but it changes. Rather than seeing the beloved directly in her physical form, one sees the beloved in the evening sky, the face of a child, in a great work of art, in the world's beauty. The love moves from particular to universal.

In the recognition that nothing in our world is immortal and that every particular thing we love will cease to exist, we experience a paralysis of the faculty of loving. This void marks the necessary distance between the "I" and the object I love. Murdoch notes that even a secularized notion of the Good maintains its sense of mystery "because of human frailty, because of the immense distance which is involved."[24] This gap is necessary for Weil because only when we allow for emptiness can God enter the space: "Grace fills empty spaces, but it can only enter where there is a void to receive it" (*NB*, 198). Through our experiences of suffering and of joy, our love can turn from natural to supernatural: "Pleasures and pains can be equally useful for wearing down the 'I'" (*NB*, 462). But because the world is bereft of God and our recognition of this requires the acceptance of mortality (and with it the inevitable loss of all particular love and even our own earthly bodily frame), suffering offers a depth of love beyond joy because it takes one through the dark night of the soul. McCullough notes that joy and suffering are similar in Weil's thought, but while joy increases one's sense of reality, suffering diminishes reality and brings about a sense of futility and total emptiness: "The beauty of the real then shines through suffering more purely than it can ever do through joy" (*RP*, 20). The sense of absence is not just between I and the other being but is also between the real and God: "Affliction makes God appear to be absent for a time, more absent than a dead man, more absent than light in the utter darkness of a cell. A kind of horror submerges the whole soul. During this absence there is nothing to love" (*WG*, 70). When we endure the absence and continue to love with a close attention, the direction of our love begins to shift. Weil tells us:

> Lovers or friends desire two things. The one is to love each other so much that they enter into each other and only make one being. The other is to love each other so much that, with half the globe between them, their union will not be diminished in the slightest degree.
>
> (*WG*, 74)

The first instance desires to subsume the particular, the second allows for the particular to exist as separate while maintaining a loving connection. Loving is charged with desire to overcome difference in the first instance of loving objects alone, but in the second instance, when one loves God, one accepts a space or a distance in one's love, which ultimately allows for difference to manifest, grace to enter, and our natural love to be given a supernatural depth.

The love of God manifests in the individual not as an exertion of the "I" but as an extension of God Itself. This depth of loving God does not relieve the pang of absence or the affliction of existence. On the contrary, it asks of us that we bear that affliction: "If we want to have a love which will protect the soul from wounds, we must love something other than God" (*NB*, 497). This posture towards God saves God from being a mere balm or escapist consolation. It also makes sense of the atheistic purification that takes place in Weil wherein one purges oneself of all such comforts in God. Meaney explains that "[t]he human heart is meant to be broken in some way. Weil hopes to inspire her readers to let God break it rather than anything else" (*AL*, 177). Esposito notes that for Weil, love "carries within itself a wound that does not heal, desire cannot be sated; its fire burns and purifies" (*OP*, 70). As such, the love of God is a continual yearning, a desire for the good that is unquenchable. Attending to the love of the particular, and the reality of absence, one moves into the love of God or the Universal Good.

Consent as an Open Mediation (Between Willing and Suffering)

The mediation of consent which brings together the faculty of willing and the necessity of suffering is best conceptualized through *amor fati*, or the Stoic love of fate. In the *Meditations*, Marcus Aurelius ties the ability to pay attention to a recognition of loving one's own fate or *amor fati*:

> Remember to retire into this little territory of thy own, and above all do not distract or strain thyself, but be free, and look at things as a man, as a human being, as a citizen, as a mortal ... Everything harmonizes with me, which is harmonious to thee, O Universe. Nothing for me is too early nor too late, which is in due time for thee. Everything is fruit to me which thy seasons bring, O Nature: from thee are all things, in thee are all things, to thee all things return.[25]

The phrase *amor fati* itself originates with Friedrich Nietzsche, but while there are signs of Nietzsche's influence on Weil, Stuart Jesson suggests that Weil's primary influence in this is the Stoics rather than Nietzsche.[26] Jesson notes that Nietzsche's ethical use of *amor fati* is found in his notion of "affirmation," whereas it is found in the notion of "consent to necessity" in Weil's work.[27] For Nietzsche, the will to power leads to *amor fati*, whereas for Weil, it is through the renunciation of the will that consent can be achieved. According to Rowan Williams, it is through *amor fati* that Weil conceptualizes God:

> The reality of God is the truth that the world can be and in some sense already is, seen and affirmed, loved as a whole; that it is possible to say yes to all that is or has been in the world, and that this possibility is entirely independent of what I or any other individual as a matter of fact can or does achieve.[28]

Against the Cartesian notion of will as that which proves our closeness to God in its limitlessness, Weil asserts that while human willpower may have "a certain sentiment of choice," she finds that it is "subject to necessity" and "admits limits" (*IC*, 181). Our will is not infinite, in that it must adhere to the laws of natural science and necessity. Necessity is bound up with will, in that it acts as both the transcendental condition for willing and the limitations of will. In other words, necessity gives the conditions upon which the will can assert itself but also provides the limits of reality to which the will must adhere. For Weil, the assertion of the will to make meaning leads to illusion because the will is controlled by earthly necessity. Weil posits the will as a vain attempt to control that which is beyond our control. Rather, the closest we can get to willing is consenting to God's Will, a variation of the Stoic *amor fati* in which we accept and even love the will of God no matter what it may bring about in our lives (*IC*, 184–185).

Weil notes that willpower can assist the human being in the embodied action of manual tasks, but when meaning creation and truth are involved, willing is an obstacle rather than an asset:

> Willpower, the kind that, if need be, makes us set our teeth and endure suffering, is the principal weapon of the apprentice engaged in manual work. But, contrary to the usual belief, it has practically no place in study. The intelligence can only be led by desire.
>
> (*WG*, 61)

The will helps to endure physical exertion and suffering, but according to Weil it has no place in higher-order practices of knowledge, spirituality, and decreation. Instead, Weil sees *attention* as key. The negative effort of attention necessary for prayer, meaning, and knowledge is the very antithesis of willpower. Whereas Weil maintains important positions for knowing and loving in her philosophy, albeit with major addendums and transformations through necessity, the will as a meaning-making faculty, as exemplified in Nietzsche and Descartes, has no space in her philosophy. Instead, Weil requires willing to shift to consenting, which is a radically different faculty. This is certainly a much starker difference than the movement from knowing to wisdom or from love of particular to love of God.

Consent, as the embrace of *amor fati* that holds open the will over and against the suffering of existence, is embraced by myriad schools of thought. Practices of mindfulness in Stoic, Nietzschean, and Buddhist schools of thought recognize that life involves a certain amount of suffering. For example, the *Four Noble Truths* of Buddhism begin with an acknowledgment of the truth of suffering (1. *Dukkha*) and it is only from the recognition of suffering that one can comprehend the origin of suffering as located in desire or attachment (2. *Samudaya*), the cessation of suffering as renouncing or letting go (3. *Nirodha*), and the path one must follow to achieve this renunciation (4. *Magga* or the noble eightfold path).[29] For Weil, inspired by this and by other traditions of mindfulness, it is the recognition of suffering over and against willing that opens up the ethical possibility of consent. Williams explains Weilian consent: "Weil has developed a scheme in which particular aims or wants are not intelligible without reference to an overarching consent to or wanting of reality as a whole."[30] The reality of the world can only be offered up when we consent to that reality and indeed, go a step further and love that reality: "the 'true' desire of the self is total consent to the ensemble of reality."[31] Weil sees this consent as "an adherence to the beauty of the world, the Stoic *amor fati*, the adherence to that indiscriminate distribution of the light and of the rain" (*IC*, 175).

Consent to reality is, for Weil, a consent to necessity: "From the time we renounce thinking in the first person, by consent to necessity, we see it from outside, beneath us for we have passed to God's side" (*IC*, 187). Weil recognizes that we cannot ever fully consent to necessity or view the world absolutely from a third-person position. Indeed, she believes that the self remains but through attentive decreation, there is a shift in the faculty of will to the open mediation of consent. Also note that one consents to *necessity* and not to *force*. While consenting to necessity, one maintains the potential to bring force into balance and push back against cycles of violence. In this way, one can blend openness with action in a movement towards ethics. McCullough notes that human effort is never a mastery for Weil, "but always actually a form of obedience" (*RP*, 32). To act ethically is to consent to and obey certain limitations but also to push beyond *ersatz* or false limitations.

Weil speaks of a fidelity to the order of the world, in which humans can use attention to notice the structure of existence mathematically, scientifically, through analogies and in relation to other beings: "The first lesson of this contemplation is not to choose but to consent impartially to the existence of all that exists. This universal consent is the same thing as detachment. And any attachment, even the weakest and most legitimate in appearance, is an obstacle to it" (*IC*, 190). For Weil, consent adheres to the creative will of God. As Susan Anima Taubes states, "To consent to affliction, Simone Weil claims, is to consent to the will of God."[32] Thus, the love of fate for Weil, is the shift from one's own will to consenting to the will of God. As such, one begins to release the ego-driven nature of the small-I as the force of willing and instead begins to recognize one's own place in the world as a part of a larger ecosystem. Rather than asserting the will upon the world, one begins to cultivate consent.

According to Weil, "The faculty which does not belong to this world is the faculty of consent" (*IC*, 182). The faculty of consent may strike us as somewhat absurd or weak at first glance, but in fact, there is an absolute fortitude in acknowledging our own limits and possibilities. This can be conceptualized through the Serenity Prayer, popularized through Alcoholics Anonymous. The prayer asks: "God grant me the serenity to accept the things I cannot change, the courage to change the things I can and the wisdom to know the difference."[33] Consent in Weil is like serenity in this prayer. It is an acknowledgment of that which is beyond one's own control and as such, is in the control of fate or God.[34] Consent makes possible the grace of God, and in this way, the transformation of willing into consenting is the transformation from *my* will to *God's* will. Weil asserts that our will is not infinite, in that it must adhere to the laws of science and necessity. Our consent, on the other hand, because it is God's will and beyond the limitations of necessity, is limitless. Consent gives us the "equilibrium between human will and necessity" (*IC*, 180).

Suffering is the catalyst for this shift in that it causes us to recognize the limitations of the will. As with the other faculties, the shift from willing/suffering into consenting may seem to be a synthesis of sorts, but ultimately each individual element remains unsynthesized, and consent is the open mediation between the two. This movement to consenting is not the end of willing altogether for Weil. Our faculty of willing is necessary in order to "un-will" itself, and this is a continual process that is Sisyphean in nature, but importantly, this is a Sisyphus that brandishes the smile posited by Albert Camus.[35] We consent willingly and sometimes even joyfully to this process, despite our inevitable suffering. McCullough notes that an

> effective way of pursuing this negation of the will without inadvertently reinforcing willfulness is to shift the will's desire away from *something* – whatever supports its substantial being – to *nothing*. That is to say, if one succeeds

> in training one's will to desire the void, the will as such is debilitated and transcended.
>
> (*RP*, 35)

Our willing, unwilling, and consenting is repeated continually. There is no end to this process as long as we are on earth. Thus, the three faculties and their necessary impossibilities to deliver on the promise of meaning and truth can be held open through attention. Productive open mediations are established in order to shift into divine ethical action. In the next chapter, I will turn to the decreative culmination of these ethical actions.

Notes

1 Sharon Cameron, "The Practice of Attention: Simone Weil's Performance of Impersonality," *Critical Inquiry* 29.2 (Winter 2003), 218.
2 Simone Kotva, *Effort and Grace: On the Spiritual Exercise of Philosophy* (London: Bloomsbury, 2020), 138.
3 See Aristotle, *Nicomachean Ethics*, trans. Terence Irwin (Cambridge: Hackett, 1999), Book VII, Chapter 7, 1150a9–14, p. 109.
4 See A. Rebecca Rozelle-Stone and Lucian Stone, *Simone Weil and Theology* (London: Bloomsbury, 2013), 95–96.
5 See Gary Watson, "Skepticism About Weakness of Will," *The Philosophical Review* 86.3 (1977), 325; Alfred R. Mele, "Self-Control, Action, and Belief," *American Philosophical Quarterly* 22.2 (April 1985), 169–175.
6 See Plato, *Symposium*, ed. M.C. Howatson and Frisbee C.C. Sheffield, trans. Howatson (Cambridge: Cambridge University Press, 2008), 198a–212d, p. 32–51.
7 Ibid., 200b, p. 35.
8 Ibid., 201b, p. 36.
9 Ibid., 202d, p. 38 and 202e, p. 39.
10 Eric O. Springsted, *Christus Mediator: Platonic Mediation in the Thought of Simone Weil* (Chico: Scholars Press, 1983), 196.
11 Ibid., 198.
12 Peter Winch, *Simone Weil: "The Just Balance"* (Cambridge: Cambridge University Press, 1989), 80.
13 Joseph K. Cosgrove, "Simone Weil's Spiritual Critique of Modern Science: An Historical-Critical Assessment." *Zygon* 43.2 (2008), 366.
14 Kotva, *Effort and Grace*, 139.
15 Diogenes Allen, "George Herbert and Simone Weil," *Religion and Literature* 17.2, *Simone Weil* (1985), 34.
16 Vance G. Morgan, *Weaving the World: Simone Weil on Science, Mathematics, and Love* (Notre Dame: Notre Dame University Press, 2005), 42.
17 Ibid., 45.
18 Cosgrove, 366.
19 Simone Weil, *Love in the Void: Where God Finds Us*, ed. Laurie Gagne (Walden: Plough Publishing House, 2018), 33.
20 Iris Murdoch, *The Sovereignty of Good* (London: Routledge, 2001), 53.
21 Ibid. While Murdoch herself notes her debt to Weil in these essays, Weil's thought also appears elsewhere and without immediate acknowledgment, especially in her literary

works. For more on this, see Donald Purcell, "Iris Murdoch's *The Green Knight* and Simone Weil," *Cahiers Simone Weil* 19.2 (1996), 225–238.

22 Joke J. Hermsen, "The Impersonal and the Other: On Simone Weil," *European Journal of Women's Studies* 6.2 (1999), 187.

23 Ibid., 188.

24 Murdoch, *Sovereignty*, 96.

25 Marcus Aurelius, *The Meditations*, trans. George Long (MIT: Online, 1994), Book 4 (online: http://classics.mit.edu/Antoninus/meditations.html).

26 Pierre Hadot, *The Inner Citadel: The* Meditations *of Marcus Aurelius* (Cambridge, MA: Harvard University Press, 1998), 143–145. Referenced in Stuart Jesson, "'The Question in Each and Every Thing': Nietzsche and Weil on Affirmation," *International Journal for Philosophy of Religion* 86.2 (2019), 132.

27 Jesson, 135 and 140.

28 Rowan Williams, "The Necessary Nonexistence of God," in *Simone Weil's Philosophy of Culture: Readings Toward a Divine Humanity*, ed. R.H. Bell (Cambridge: Cambridge University Press, 1993), 56.

29 Dalai Lama XIV, *The Four Noble Truths: Fundamentals of Buddhist Teachings*, trans. Geshe Thupten Jinpa, ed. Dominique Side (London: Thorsons, 1997).

30 Rowan Williams, "Review of *The Just Balance* by Peter Winch," *Philosophical Investigations* 14.2 (1991), 162.

31 Ibid., 163.

32 Susan Anima Taubes, "The Absent God," *The Journal of Religion* 35.1 (1955), 11.

33 This prayer is commonly attributed to Reinhold Niebuhr. For more, see Laurie Goodstein, "Serenity Prayer Faces Challenge on Authorship," *New York Times* (July 11, 2008) (online: www.nytimes.com/2008/07/11/world/americas/11iht-prayer.4.14439421.html).

34 McCullough notes that human effort for Weil is never a mastery "but always actually a form of obedience" (*RP*, 32).

35 Albert Camus, *The Myth of Sisyphus* (London: Penguin Books, 2005), 119.

5 Decreation and Action

Simone Weil considers philosophy to be, first and foremost, action (*FLN*, x). If a treatise is thought or theory alone, it is not philosophy for Weil. She is concerned with a philosophy focused on the life lived and acted. While she wrote ceaselessly, ordering her ideas and exploring the works of other philosophers, she also lived those ideas and had very little patience for those who did not. During her time at the École Normale Supérieure, her uncompromising embodiment of ethics earned her the nickname "the categorical imperative in skirts," referring to her motivation to uphold Kant's "kingdom of ends" in which one must act in such a way that if everyone acted thus, the world would be a better place.[1] Furthermore, her ethical conviction makes sense of her affinity for the character Antigone, who symbolizes an uncompromising ethical sensibility that chooses the supernatural "higher" law over the human-decreed natural law of state power. In a letter, Weil refers to herself as "Antigone as usual."[2] For Weil, ethics are exemplified in Antigone's decision to give her brother Polyneices a proper burial despite King Creon's decree that Polyneices' body be left out in the elements without ritual burial. As she is marched to her own death, a righteous Antigone asks, "What divine justice have I disobeyed?"[3] She lives and dies by a higher law, and this is precisely what Weil seeks to philosophize and to live.

Weil's distain of philosophers who are unwilling to move beyond the armchair is demonstrated in her implicit link between the human faculties of knowing, loving, and willing and the pursuits or actions of science, art, and work. For Weil, the three faculties cannot move beyond the nihilism of absurdity, absence, and suffering just to halt upon the bridges of wisdom, loving God, and consenting, but must use these bridges to cross over into action. Lissa McCullough argues that for Weil, "care for creatures is the verifiable expression of love of God. What is good in a good act is that, in it, love acts" (*RP*, 83).[4] While I specify love particularly as the faculty behind art in this section, all actions are saturated with divine love when they are the products of one's decreation. Following Iris Murdoch's adaptation of Weil, we can see this as a necessity for moral philosophy: "We need a moral philosophy in which the concept of love, so rarely mentioned now by philosophers, can once again be made central."[5] Murdoch feels that this use of love is necessary to displace the Cartesian notion of will that creates the moral subject in the postwar

DOI: 10.4324/9781003449621-7

Oxbridge school of moral thought: "Let me now simply suggest ways in which I take the prevalent and popular picture to be unrealistic. In doing this my debt to Simone Weil will become evident."[6] Murdoch's notion of "unselfing" is influenced by Weil's decreation and offers a secularized interpretation of decreation that maintains the mystical and metaphysical style of thinking. Murdoch emphasizes the beauty of nature met with human attention in her example of seeing a kestrel.[7]

The shift from theory to action is a part of the process that Weil refers to as decreation. Action marks the second, creative movement of decreation. It is helpful to recall the distinction made by A. Rebecca Rozelle-Stone and Lucian Stone of (1) "the first movement of grace-Death, or the decreative event" and (2) "The second movement of grace-Recreation, or subjectivizing inspiration."[8] As McCullough observes, "Decreation of the soul is not the completion of our earthly task. On the contrary, our true earthly task begins in earnest with decreation" (*RP*, 196). This reading of decreation as only the act of self-abnegation is distinct from my own reading, in which I have noted decreation as necessarily including first self-abnegation to clear a space for ethical action, and then ethical action that reasserts the self as relational at the site of the other. This is not an act of will, but an open response/inspiration. With this in mind, I will begin this chapter on decreation and action by setting out the decreation of God, the individual act of decreation (as opposed to destruction), and then move into the ethical actions of science, art, and work.

God's Decreation

The original act of decreation is the creation of the universe by God. The act of creation is God recoiling Itself in order to create a space for the universe: "God could create only by hiding himself. Otherwise there would be nothing but himself" (*FLN*, 85). In this decreative act, God allows life as we know it to exist by removing Itself and thus creating time and space:

> For God, the creation consisted not in extending himself but in withdrawing. He refrained from "commanding wherever he had the power" ... The creation, the passion, the Eucharist – always the same movement of withdrawal. This movement is love.
>
> (*FLN*, 81)

Thus, in creating the world, God's action removes God and God's goodness from the world. In this sense, God's creation of the world contains within it the original sin or the distance between God and humanity that is customarily blamed on Eve and Adam in biblical tradition. In Weil's reading, God's creation contains both sin *and* salvation: It is the crucifixion, or God's sacrifice and God's purposeful rending of Itself for the sake of humanity. This sin and evil manifested in creation is explained in Weil's thought as a dialectic in which only through suffering and chaos can humans and God achieve an absolute perfection and love.

Individual Decreation Versus Destruction

A mirroring of God's creation of the universe, the individual act of decreation is a form of self-abnegation or of releasing the ego that assumes the self to be radically individual and separate. Weil notes, "We participate in the creation of the world by decreating ourselves" (*NB*, 309). Roberto Esposito reads individual decreation as a form of internalized agonism: "the assumption within the self of the firmest agon, of perfect agony not limited to a moment but extended to the totality of time" (*OP*, 72). The passive effort of attention cultivates this decreative activity. Decreation recognizes individual autonomy but chooses interconnection:

> God has given me being in order that I may give it back to him. It is like one of those tests that resemble traps and are to be found in fairy tales or stories about initiation. If I accept this gift, it has a bad and fatal effect. Its virtue becomes apparent through a refusal. God permits me to exist while being other than He. It is up to me to refuse this permission.
>
> (*NB*, 484–485)

As noted by Esposito, the decreative process is incredibly similar to the Kabbalistic notion of *tzimtzum*, although this is a similarity that Weil herself never acknowledges and it is made all the more problematic by her indifference to Judaism and even anti-Semitism.[9] Isaac Luria's theory of *tzimtzum* is described by Ann Rabinowitz as "a Kabbalistic concept which portrays God as inhaling at the moment of creation, diminishing himself to make room for the universe."[10] This means that while there is a recognition of God in the earth, the earth is not God Itself. McCullough explains that the infinite distance between God and humanity prohibits pantheism in both kabbalah and decreation (*RP*, 91). There is a desire for God in living things, a recognition of God's fingerprints or work in creating the world, but the absence of God means that at most, earthly bodies participate in a desire for the divine but are not themselves the divine.

At times Weil explains the act of decreation in ethically problematic ways that have lead critics to note a self-annihilation and a hatred of the material realm rather than a self-abnegation and mediation between matter and the Good. For example, Francine du Plessix Gray bluntly states, "There are many other aspects of Weil's thought that make me want to throw a book at her, such as her Gnostic view that the physical world, particularly the body, is the source of all evil (a belief she practiced in real life by gradually destroying her own body)."[11] Judith Gregory demands of Weil, "I want to know: what happened? How did it come about that you felt you must choose, and elaborate, and live a spirituality of self-annihilation?"[12] Claire Wolfteich argues that "Simone Weil was a woman of hunger. She yearned for God, and her intense longing for the absolute made her relationship to matter problematic."[13] Even setting attacks upon Weil's personal illness aside, decreation as a destruction is often cited as philosophically present in some of her writings. Rowan Williams notes that one's experience of God's love is contingent upon one's decreation. According to Williams, Weil's philosophy suggests that "God is loved to the extent to which creation is destroyed."[14] Williams draws out a source of tension in Weil's work between her ethically fecund concept of uncontrollable otherness and her "sense that the necessarily frustrated subject, the limited point of view, is somehow the source of error, a corruption of some potentially divine subjectivity only thinkable for us in terms of negation, passivity, absence, death."[15] Williams herein articulates a frequent critique of Weil's decreation: her ethics of otherness seems to come at the expense of the existence of the I. The I must be decreated to allow for other bodies and God to have space for ethical existence. Yoon Sook Cha examines the problematic nature of Weil's ethical claim about the other: "The cry of destitution [from the other] corresponds to the destruction of the 'I' from without" (*DEB*, 31). The external force of the other's needs necessitates the ethical response of the I as an annihilation of the self. In this critique, ethics for the other comes at the expense of one's own well-being or ethical import. The renunciation of the self notes the site of the "subject" (and by extension the site of ethical action) is not within the I nor within the Other but in the liminal *metaxu* or the space between the I and the other.

My reading of Weil's decreation as a flowing movement between faculty, necessity, mediation, and action is, in part, meant as a way to push back against the critique of Weil's decreative ethics as self-annihilation in that it offers a unique twofold approach to the decreative process that necessarily includes both self-abnegation and recreation of the self as relational at the site of the encounter with the other. Rather than a self-annihilation, her ethics is a recognition of the world of the other person and the desire for the Good that lives in that person. It is a movement in-between I and other. However, there are times when Weil's decreation does seem closer to a self-destruction. An example is the following passage:

> To empty oneself of the world. To take upon oneself the character of a slave. To reduce ourselves to the point one occupies in space and time. To become nothing … One must strip oneself of the imaginary sovereignty of the world, in order to reduce oneself to the point one occupies in space and time. Absolute solitude. Then one is in possession of the truth of the world.
>
> (*NB*, 212–213)

The first movement of decreation as a self-abnegation at times suggests a terrifying lack of care for the self. In examining this tension, one may consider that due to her lifelong poor health, Weil was unable to eat at the end of her life and died of inanition.[16] Another troubling passage arises from the final pages of her notebooks:

> One must remember that a plant lives by light and water, not by light alone. So it would be a mistake to count upon grace alone. Energy from this world is also needed … But when one is entirely deprived of this world's energy one dies. So long as my heart, my lungs, and my limbs can function at all is experimental proof that there is still a drop of water on the stone to nourish heavenly wheat … Make sure it gets water, even if this means the death of the flesh from inanition … If only my flesh and blood are dried up before the divine stalk, nothing else

> matters ... But where to find the courage to deprive one's flesh and blood of the last drop of water and give it to the divine stalk?
>
> (*FLN*, 348–349)

These words from her sickbed could suggest an annihilation of the physical body so that the soul may join with the divine. Let us briefly consider this possibility through a look at her biography. Weil pushed herself towards modest sustenance throughout her life. Her moderate consumption was certainly a lived manifestation of her philosophy, but we are not justified in conflating this with her untimely death or her health issues. Following both Weil's own testimony about the nature of decreation and that of those who knew her best, her death, while brought on by a lack of nourishment, was not an attempt to decreate her own physical form but rather the consequence of long and painful bouts of headaches and digestive ailments that had afflicted her since childhood. Simone Pétrement explains Weil's digestive problems in her final months, and it is clear that she does not believe Weil's death to be a suicide.[17] Pétrement also describes that Weil was plagued with headaches throughout her life and that often the headaches would be so severe that Weil was unable to eat: "when she had her headaches Simone would at times go for five or six days without being able to eat. She could only bear to eat raw, grated potatoes. These headaches would sometimes make her vomit; indeed, any effort, such as trying to masticate, could cause this."[18] With this, we have biographical information of a tendency to eschew food when sick, and a near-lifelong history of debilitating poor health. Philosophically, it is important to note that Weil explicitly states decreation is radically different than any form of destruction or death drive (*NB*, 247–248). However, as I have quoted above, some of her writings are admittedly less clear on this distinction. Furthermore, we cannot know if she would have stood behind any of these statements had she recovered or if they were meant only for her private notebooks. This problem spans the Weilian oeuvre because she published so little in her short life but had such extensive notebooks.

Richard Rees writes, Weil "believed suffering should be avoided, when legitimately possible, but she knew it was often unavoidable and she knew its potential value" (*FLN*, viii). There is an important distinction to be made between the suffering that is an inevitable aspect of human existence (namely, the fact that all material things wield force, are subjected to force, and will inevitably die) and self-inflicted suffering. Weil notes that we must not seek out affliction and if we do, it will lose its supernatural potency. Esposito explains decreation: "This does not mean that the subject should disappear, but that subjective intentions or declinations should remain on the wane"[19] (*OP*, xiv). With this in mind, purposeful starvation would lose any spiritual potency, according to Weil's own thought, in that one must not seek out or self-inflict affliction. With this in mind, McCullough emphasizes the distinction between decreation and suicide:

> Suicide is an act of the will, whereas decreation – in diametric contrast – is an act of supernatural love responding to grace. Suicide is condemned as an "ersatz form of decreation" (N 262). We cannot effect this death of the Pauline "old

> man" ourselves (N 78–79); we have got to be killed as an effect of grace, not will. "We must die – not commit suicide, die, be killed, not literally, but nearly so, feel through the fact of external things, the chill of death" (N 61). Suffering and affliction communicate this chill of death, which if it were not pressed upon us by necessity, we would never actually face or undergo.
>
> (*RP*, 186)

Taking into account the biographical, philosophical, and ethical implications, it becomes increasingly unlikely (although admittedly not impossible) that Weil's decreation supports or advocates a self-destruction. In the reading I would like to suggest, a reading which aligns with much of Weil's own writing excluding the writings during her final illness, Weil's decreation is not to be understood as a destruction of the physical or even spiritual body but rather a preliminary step in an ethical process that expands and reimagines the self in relation to others.

Cha notes that "decreating the 'I' must be understood as a creative act, one that might even be said to participate in divine creation" (*DEB*, 36). This suggestion is an important counter to the idea of decreation as destruction and it is absolutely necessary for the sake of a possible ethics in Weil. However, Cha also acknowledges that "[i]t is hard to read these statements without the suspicion that self-effacement is not also an act of massive self-violence" (*DEB*, 37). Rather than shying away from either decreation's self-destruction *or* its creative potential, I suggest we look to how Weil keeps both creation and destruction open and pinpoint the times that she perhaps fails in her own writing and life to strike the balance towards which she otherwise so adamantly strives. Cha's relational reading of decreation as a continual process offers a counter to reading decreation in Weil's last writings as an absolute self-dissolution. Here it is important to emphasize "the overcoming of the individual perspective, not by the denial or dissolution of points of view, but by a consistently relational and dialogical account of the shaping of human awareness,"[20] which Williams sees as possible within Weil's work. This is not always the path of Weil's notebooks, particularly in her final throes of illness, but it is dominant in her major later works and her reading of Plato (*NB*, 145, 440).

Weil likens the act of decreation to Buddhist *detachment* and to the Stoic *amor fati*, love of fate (*NB*, 550). By embracing the void, suffering, death, absence of existence, by stepping into the void with no hope of redemption, only then can one experience God (*NB*, 137–138): "Those who persevere in love hear this note from the very lowest depths into which affliction has thrust them. From that moment they can no longer have any doubt" (*WG*, 72). The decreated movement is described by Gustave Thibon: "To make something created pass into the uncreated" (*GG*, 78).[21] Weil is careful to distinguish decreation from destruction, which is to pass from the created into nothingness (death, nihilism, or the lower void). The uncreated, however, is the realm of God and the goal of decreation (*NB*, 247–248). The passive action of removing oneself is what Weil calls love: it withdraws the self. This is the power behind the three actions that uphold the faculties: they allow for a creative, interconnected, and relational decreation.

As self-abnegation, decreation is not an act of the will. While one must cultivate or prepare oneself for decreation, one cannot simply *will* it to happen. Lawrence Schmidt notes that while we may see ourselves as enacting decreation, the process happens to us whether we choose it or not:

> To be decreated does not mean that we are annihilated but that we are diminished into God, diminished as fragile bodies, as anguished souls and as social beings starved for recognition. We cannot decreate ourselves. It happens. Disappointments and failures, betrayals and inattention, spiritual emptiness in the presence of material abundance, crushed or crashing careers, family members unworthy of our sensitivity and care, sickness and aging in spite of our exercise programs and vitamin treatments, and finally death which does have dominion – all these strip us of our individuality. And we resist but must finally submit to the renunciation of self.[22]

One cannot prevent decreation, or the renunciation of the I from happening over time. While Schmidt argues that we cannot decreate ourselves, I would add to this that we cannot decreate ourselves without engaging in the affairs of the world. I believe one can choose to decreate in the face of adversity. One will be decreated regardless, but one can choose to allow one's own suffering to be an opportunity for growth and connection. Weil argues that one cannot decreate oneself in solipsism (one needs the trials and tribulations of the external world), but one must participate in this process in order to learn and grow. One must cultivate attention and seek to continually balance the self in the flux of the world. Weil does not want the decreative process to happen to the individual upon death (although, as Schmidt states above, it will be forced at that time if one has not done the work already). Rather, one should work on decreation throughout one's life as the Buddhist works on detachment in an effort to live ethically and die truthfully. The actions of science, art, and work are also referred to by Weil as "creations" in that she sees them as the ways in which humans can re-create the world: "By work, he [the human] creates his own natural existence. By science, he re-creates the universe through symbols. By art, he re-creates the alliance between his body and his soul" (*FLN*, 44). These acts are the re-creation portion of the decreative process, and they allow the person who has shed her ego (in a mirroring of God's decreation) to help create a just world.

The Action of Science (via Wisdom)

Arising from the faculty of knowing, the necessity of absurdity, and the mediation of wisdom is the action of science: "Harmony and presence in the sensible world of mathematical necessity proportionate to the limits of our understanding ... With respect to beauty, necessity, in mathematics, holds the place of matter" (*NB*, 513). Namely, science, which includes mathematics for Weil, is the study of harmony in the material world. According to Thibon's editing of Weil, "The subject of science is the beautiful (that is to say order, proportion, harmony) in so far as it is suprasensible and necessary" (*GG*, 204). To do science is to put our knowing to use in the world: to find order, proportion, and harmony in our surroundings while acknowledging the absurdity of our existence. Marie Meaney explains that Weil attempts "to renovate science by reintroducing the Greeks' understanding of it. Search for truth should be its defining characteristic, rather than the modern pragmatic approach that thinks only in terms of hypotheses" (*AL*, 53). Great science holds absurdity in the balance and investigates the wonders of the material world: "The object of science is the presence of Wisdom in the universe, Wisdom of which we are the brothers, the presence of Christ, expressed through matter which constitutes the world" (*WG*, 108). Science draws out the beauty of the world, finding order and, as such, the trace of God in a world almost devoid of God. Vance Morgan notes that the proper outcomes of Weilian science are never mere matters of fact or relations of ideas in a positivist sense, but rather "order, necessity, equilibrium, beauty, truth, and justice."[23]

In both the scientist's process as well as the observation and use of science by laypersons, a potential *metaxu* (or link to the supernatural via the natural) emerges because it necessitates attention to the natural world and the supernatural order of the world. This scientific *metaxu* may be found in any number of investigations, for example, the study of a Nautilus shell and its relation to the golden ratio as symbolically represented by φ (*phi*), where $\varphi = (1 + \sqrt{5})/2 \approx 1.618$, or perhaps to the meta-golden chi ratio of 1.356, or, as has been argued by Christopher Bartlett, "One should keep in mind that each Nautilus specimen is a natural form with inherent biological variability and will not conform perfectly to strict mathematical parameters … There are also intraspecies and individual specimen deviations … The Nautilus simply follows its genetic imperative growing gracefully larger without changing shape."[24] The possible *metaxu* here is not derived from the perfect correlation of the golden ratio and the Nautilus shell, because indeed, it has been proven that the shell is not perfectly aligned to either the golden ratio or the meta-golden ratio. Rather, it is the attention and study, the investigation and openness to the shell that invites the possibility of *metaxu*.

Morgan notes that contemporary science is so problematic for Weil because it seeks to solve fundamental contradictions: "The artificial, reductionistic imposition of unity of contradictory phenomena requires a continuous denial of that feature of human experience that most directly turns our attention to value and goodness."[25] According to Weil, math is the science of nature par excellence, and she sees its zenith in the Pythagoreans:

> [In math] one can conceive an order of certainty, starting from uncertain and easily grasped thoughts about the sensible world, proceeding to thoughts of God which are absolutely certain and absolutely inapprehensible … It is easily understood that when they perceive this poetry, the Greeks were intoxicated by it; they had the right to see in it a revelation.
>
> (*IC*, 164–165)

Weil's understanding of science as the product of attention to the natural and a grace-filled openness to the supernatural perhaps brings it closer to a contemporary view of art than of science. This reading of science is aligned with Plato's theory of the Forms and humans' epistemic return to the realm of the Good through attention to beauty, symmetry, and the geometry of the material realm. Morgan notes that this understanding of the world as a Pythagorean *kosmos*, or a beautiful ordering, "combines the notions of regularity, tidiness, and arrangement on the one hand with beauty, perfection, and positive moral value on the other."[26] For Weil, as for the Pythagoreans, both art and science are actions derived from the possibility of *metaxu*, or the mixture of natural and supernatural within an object:

> There is a higher degree of musical technique in Gregorian plainsong than in Bach and Mozart themselves; the Gregorian chant is at once pure technique and pure love, in which, moreover, it is like all great art. It must be exactly the same for science, which, like art, is nothing but a special reflection of the beauty of the world.
>
> (*IC*, 172)

The likeness of science and art for Weil links them as actions in the embodied experience of decreation.

Weil's mediated understanding of science is a far cry from the positivist objective science described in the initial assertion of an *ersatz* faculty of knowing. It is also distinct from the contemporary understanding of western science, which Weil sees as having predominately lost its purpose. Weilian science is an act of true attention, which is in absolute opposition to the practice of science today. In considering Jane Bennett's use of attention, Simone Kotva notes, "Attention as it is used today in science and philosophy is good for isolating the environment as an object 'over there' but it is not attentive and remains insensitive to the 'everyday experience of … comingling' between selves and environment that attentiveness brings to light."[27]

In a Platonic understanding of science, Weil interprets the possibility of science as a way to see the Good at play in the world, and as such a way to understand how to act morally in the world. As an example of how this moral dimension acts as a foundation for scientific action, consider Kamran Loghman, co-developer of pepper spray, who noted "I have never seen such an inappropriate and improper use of chemical agents,"[28] after witnessing the police pepper-spraying students demonstrating against tuition hikes at University of California Davis in conjunction with the Occupy movement. Similarly, Albert Einstein expressed deep regrets that his own scientific discoveries led to the creation of the atomic bomb (although he did not himself work on said bomb's creation).[29] Even J. Robert Oppenheimer, the head of the Manhattan Project and the "father of the atomic bomb," expressed complicated regrets about his role in creating the atomic bomb.[30] The staggering regret of physicists and scientists working in nuclear or chemical fields whose discoveries lead to human atrocities would be averted in this mediated science, in that the exploration of knowledge is tempered with the absurdity and force of existence. Thus, in wisdom, the scientist seeks the Good through which humans can be protected and offered dignity. Within this mode of scientific exploration, the creation of the atomic bomb is completely non-sensible. Morgan notes Weil's *The Need for Roots*, in which Weil "includes a call for a radically changed vision of science, rooted in a sacramental conception of reality that alone, in her estimation, can renew hope for humanity in the wake of the tragedies of the first half of the twentieth century."[31] Such a science upholds a critical ethical commitment first and foremost.

The Action of Art (via Loving God)

Arising from the faculty of loving, the necessity of absence, and the open mediation of loving God is the action of art: "Art is an attempt to transport into a limited quantity of matter, modeled by man, an image of the infinite beauty of the entire universe. If the attempt succeeds, this portion of matter should not hide the universe, but on the contrary should reveal its reality to all around" (*WG*, 107). Artistic genius for Weil is a phenomenological ability to pay attention. In this way, artwork recognizes the absence of God and the violent force of existence and marries it with the beauty of the world, which Weil sees as the trace of God in our surroundings. Murdoch argues that "Beauty is that which attracts this particular sort of unselfish attention."[32] She claims that beauty is that which demands decreation and attention. Furthermore, the proof of the Good for Murdoch can be found in experiences of art.[33] Weil argues that great art channels the force of gravity and enacts it with a love of the world's beauty: "Music. Are the opposites represented by the ascending and descending movements? It is a question of rendering the descent to the second degree sensible – that descent which represents love and not gravitational force?" (*NB*, 486). This question is put in stronger language in Thibon's edited version: "A double movement of descent: to do again, out of love, what gravity does. Is not the double movement of descent the key to all art? This movement of descent, the mirror of grace, is the essence of all music" (*GG*, 206). While science is attention to a pure alchemy of beauty and order for Weil, art also plunges into descent, gravity, force in order to repeat that descent with beauty and love.

All great art, for Weil, is sacred because it is beautiful: "In everything which rouses in us a pure genuine feeling for beauty, God is really and truly present. There is, as it were, a sort of incarnation of God in the world (*Timaeus*) of which beauty constitutes the sign" (*NB*, 440). The world is decreated and thus devoid of God and yet in the beauty of the world, Weil finds an incarnation of God. As McCullough notes, for Weil "Saintly action incarnates God in the world" (*RP*, 204). When our actions are ethical, God manifests on earth. In this way, God's trace is momentarily captured by great art. The action of art is one of attention and of bringing supernatural prayer into natural matter.

Again, we see here Rozelle-Stone's second movement of grace or my second movement of decreation: a re-creation that has been through the dark night of the soul, looked into the void, and has chosen to love despite the erosion of worldly meaning. As God creates the world by retracting Itself, we can create art by retracting ourselves and allowing for a mimesis of God's creation of the world: "In art and science of the 1st order, creation is self-renunciation" (*FLN*, 93). In viewing art, we must take up the aesthetic sensibility of seeing rather than eating: "It may be that vice, depravity, and crime are nearly always, or even perhaps always, in their essence, attempts to eat beauty, to eat what we should only look at" (*WG*, 105). Art is to be taken in but not to be wholly consumed. One must allow for the art to exist in itself and not to be totalized into one's own perspective. In this way, art can act as *metaxu*: "It is the triumph of art to lead to something other than itself: to a life which is fully conscious of the pact between the mind and the world … art is exploration" (*FLN*, 44). Her own close readings of ancient Greek tragedies showcase her belief that art can reveal to us the void, the beauty of existence, and our ethical duties. McCullough notes that for Weil tragic art "embodies the essential contradiction of creaturely existence – that life is impossible, that existence per se is the crucifixion of infinite love by inescapable necessity" (*RP*, 142). Art recognizes the impossibility of fixed meaning and the cruel necessities of earthly existence, it pays close attention, and it traces these cruel necessities with love, presenting them as painting, poetry, music, theater, dance, or another artistic expression. It recognizes the work of gravity and repeats that work infused with attentive love.

The Action of Work (via Consent)

While art and science relate to beauty for Weil, work is categorically deprived of beauty. Work is beauty's inverse:

> A squirrel revolving in its cage and the rotation of the celestial sphere: on the one hand, supreme wretchedness; on the other, supreme splendour. It is just when man sees himself as a squirrel revolving in a cage that, provided he doesn't lie to himself, he is close to salvation.
>
> (*NB*, 496)

Work is closer to divinity than art or science, but particularly in industrial society, it is also worlds away because it enslaves the worker for no reason or good other than maintaining her continued mere existence as a commodity. The worker lives in a parallel existence to divinity but because this stripping of identity is not by free choice and because the worker lacks attention and must detach herself from the process of her work in order to survive, it seems almost impossible to move into the higher void: "Work is like a death if it is without an incentive. We have to act, renouncing the fruits of action" (*NB*, 78). Here, Weil is echoing the famous verse from the *Bhagavad Gita* that says "You have the right to work, but for the work's sake only. You have no right to the fruits of work. Desire for the fruits of work must never be your motive in working."[34] Weil points to work as necessarily about the action and not the action's fruits. Industrial laborers, however, are robbed of this in that they cannot *choose* to renounce the fruits of their labors. As such, it becomes the ethical responsibility and the goal of decreation to allow the other to exist without affliction imposed upon them in order to merely survive. All that oppressed workers can labor towards is mere existence. They do not have the luxury of working towards a higher good. But as in the enlightenment of the Gita, the ultimate act of decreation comes about through work in that it strips the subject of the identity to which she clings.

The potential decreative process of work can be found in practices of service that are voluntary and not meant to merely maintain a bare life. The work of religious orders such as the practice of *seva* in ashrams allows for decreation in that the devotees spend their days doing physical labor that will exhaust them so thoroughly that they have no choice but to let go of their ideas about who they are, what they can or cannot do, what they are above, and what is below them. But the worker in industrial society often remains trapped in the slough of labor without the epistemic and ontological structure that offers decreative benefits: "As God has created our independence so that we should have the possibility of renouncing it out of love, we should for the same reason wish to preserve the independence of our fellows" (*WG*, 116). Kotva points to the difference that Weil sets out between a pregnant woman sewing for her child and a prisoner sewing in a prison workshop. The women are both technically doing the same work, but "the pregnant woman's attention is motivated by love for her baby while the convict's attention is motivated by fear of her foreman."[35] The point is to allow for free attention motivated by love and to do our utmost to eradicate attention motivated by fear. Ranilo Hermida notes that it is only the free human who can decreate, and as such, we have an ethical responsibility to free the other from oppressive material enslavement and forced destruction of the I so that the other can choose decreation freely.[36] The work that Weil herself embodied is noted by Benjamin Davis as bringing her theories into the world: "she wrote to clarify the wishes and the struggles of her age, but she never stopped there. Going further, she tested her ideas through practice – on factory floors, in the Spanish Civil War, in Marseilles courtrooms, alongside workers in Harlem, and in dialogue with other exiles in London."[37] Work, for Weil is the culmination of her philosophical method.

The faculty of willing, the necessity of suffering, and the open mediation of consenting manifest the action of work, which is the physical laboring that consents to God's will and seeks to manifest it in the world. While knowing and loving create the mediations of wisdom and loving God as openings into beauty, working opens us up to God or the ethical responsibility of the Good through suffering and affliction. The most potent of the three actions and the one favored by Weil, work's reliance on suffering contains within it the desire for joy. Indeed, Weil sees joy and suffering not as opposite but as agonistically connected: "Thus suffering in affliction and compassion for others are all the purer and more profound the better we are able to conceive the fullness of joy. Of what does suffering deprive him who is without joy?" (*NB*, 290). Weil argues that ethical action is contingent on comprehending both joy (that which is precious and at stake in our worldly existence) and suffering (that which is capable of destroying what is precious). As such, work becomes richer alongside the joy and beauty of art and science.

Weilian Ethics

While each of these actions is individual, Weil believes that ultimately, they must be enacted together: "Note that each of these 3 creations [actions] is a poor, vain, empty thing when taken by itself and unrelated to the other two" (*FLN*, 44). The three actions must co-exist if the individual is to enact decreation, shed her sense of absolute autonomy, and comprehend how to act ethically. Pétrement explains that Weil's is an "action that is divided, an action of parts at once tied together and distinct, tied together but indifferent to each other, inseparable and separable. And that signifies, for Simone, that work implies an extended world."[38] All three of the faculties must be cultivated into these actions if one is to be able to perceive her ethical duty, which at its heart, is born in the moments when others have spared her from the force they might have inflicted upon her. This demands one to balance force in her own life. Cha calls this the *ethical bind to the other* in which my own subjectivity is renounced for the sake of the other and as such I am exposed to "vulnerability, failure, even death" (*DEB*, 39). The ethical bind entails that

> bound to the other by an unconditional and unconditioned demand ("Why am I being harmed?"), one must nonetheless undergo and sustain an unconditioned desire defined by its unrealization and that presumes the destruction of one's "I." If we follow Weil, it is the only thing that one can offer.
>
> (*DEB*, 39)

McCullough notes that "To live and work continually inspired by a consciousness of endlessness is to convert all labor and indeed all action into sacred ceremony" (*RP*, 212). To have power and to choose not to inflict it upon the other because of my own gratitude and love of others who have done the same for me is a miracle for Weil. This miracle, which reveals a divine goodness in a world of inordinate existential pain, is built into the core faculties of our human condition and as such, requires the recognition that our thoughts and emotions are intertwined with our actions.

The faculties that belong to each person only remain useful in our search for the truth if we use them to cultivate attention, decreation, and the ethical space of the other. Truth, for Weil, is a higher good than God, although she believes that while it may take us through atheistic detours, the search for truth ultimately brings us back to God. Discussing Emmanuel Levinas' reading of Weil, Emmanuel Gabellieri notes that what Levinas understood in Weil was "the positions they shared: first, an openness to an alterity that rejects the ontology of the same … second, and more significantly, the similarity of his theme of the substitution of 'the One for the Other' and the rending of the self in Weil's concept of decreation."[39] The decreative act is the act that allows for the Other. It is an ethical mediation, which Gabellieri points to as Weil's adaptation from Plato.[40] The truth that Weil locates within Plato and takes up in her own philosophy is a movement of relation and a continual process of decreation that allows for the existence of the other in order to move towards the Good. In asserting the role of God in this equation, Williams claims that decreation is a way in which we refine our love:

> to purify love is to learn how egoistic fear and fiction work to smooth out the particular Otherness of another person, so that my language remains uninterrupted, my control unchallenged, my involvement in time and chance unacknowledged. And to know this contingency in the event of love is precisely to retain and nurture an apprehension of the difference of this or that other, their own contingency; to be surprised, delighted, puzzled, hurt by them in a way which witnesses to their un-assimilated reality, an independent hinterland to their side of the conversation.[41]

This decreative movement is a movement towards God and towards the vital experience of other people as they are capable of forming my world. Ethical actions are embedded as seeds of desire for the Good within our human faculties. Through the decreative process, these seeds can grow. This process is never completed (until perhaps our physical death) but is an ongoing negotiation. As such, we can never reach a state of perfection or pure decreation but we must continually strive towards a balance between natural and supernatural, bringing justice to our action.

The relational and continual movement of decreation can be cultivated through our attention to the faculties of knowing, loving, and willing; our recognition of the impossibilities of these faculties; and our worldly actions that uphold faculty and impossibility as we actively work towards decreation and ethical being through science, art, and work. In this way, Weil binds together theory and praxis and posits an ethics founded on our responsibility to the other and realized through this process of decreation. In various places I have suggested the radicality and importance of Weil's interpretation of Plato and I will turn to this in greater depth over the next two chapters, which will situate how and why Weil's interpretation of Plato is so important to her work, her place in the larger philosophical canon, and her relevance for the ecological ethics that I will unpack in the third and final part of this book. Meaney writes that Weil "obviously did not just want to present some new ideas, but to bring about a change of heart" (*AL*, 111), and Davis notes "Weil's method as twofold, consisting in understanding and action or, more specifically, in inquiring and essaying."[42] Weil's work demands far more than a mere intellectual inquiry. It takes philosophy as the fierce whisper from the Archaic torso of Apollo: "You must change your life!"[43]

Notes

1 Jacques Cabaud, *Simone Weil: Fellowship in Love* (London: Harvill Press, 1964), 36.

2 Simone Weil, *Seventy Letters*, trans. Richard Rees (Oregon: Wipf and Stock, 1969), 161. For more, see Simone Weil, "Antigone" in *IC*, 18–23.

3 Sophocles, *Antigone*, trans. Elizabeth Wyckoff, in *Greek Tragedies I*, 187–240, ed. David Grene and Richard Lattimore. Third edition ed. Mark Griffith and Glenn W. Most (Chicago: University of Chicago Press, 2013), line 920, p. 225.

4 For Weil, "creature" refers to humans in their distinction from the creator.
5 Iris Murdoch, *The Sovereignty of Good* (London: Routledge, 2001), 45.
6 Ibid., 49.
7 Ibid., 82.
8 A. Rebecca Rozelle-Stone and Lucian Stone, *Simone Weil and Theology* (London: Bloomsbury, 2013), 165–171.
9 Rozelle-Stone and Stone offer a synopsis of texts covering Weil's anti-Semitism. See Rozelle-Stone and Stone, *Simone Weil and Theology*, 51–66.
10 Ann Rabinowitz, "Tzimtzum," *Judaism: A Quarterly Journal of Jewish Life and Thought* 47.2 (1998), 213.
11 Francine du Plessix Gray, "At Large and at Small: Loving and Hating Simone Weil," *The American Scholar* 70.3 (2001), 8. www.jstor.org/stable/41213171.
12 Judith Gregory, "A Letter to Simone Weil," *Cross Currents* 40.3 (1990), 368. www.jstor.org/stable/24459577.
13 Claire Wolfteich, "Attention or Destruction: Simone Weil and the Paradox of the Eucharist," *The Journal of Religion* 81.3 (2001), 359. www.jstor.org/stable/1206400.
14 Ibid.
15 Rowan Williams, "The Necessary Nonexistence of God," in *Simone Weil's Philosophy of Culture: Readings Toward a Divine Humanity*, ed. R.H. Bell (Cambridge: Cambridge University Press, 1993), 72.
16 This was the official cause of death for Weil and is defined as a state of emptiness or an exhaustion from lack of food and water.
17 Simone Pétrement, *Simone Weil: A Life*, trans. Raymond Rosenthal (New York: Pantheon Books, 1976), 527–528.
18 Ibid., 70 and 82.
19 Esposito, *OP*, xiv.
20 Williams, "Necessary," 73.
21 While this definition is accredited to Weil in *Gravity and Grace*, it is not found in her notebooks or essays and is perhaps Thibon's interpretation of Weil rather than her own words. That being said, it seems to me to be a tenable interpretation of her decreation.
22 Lawrence Schmidt, "The Language of Limitation as the Key to Simone Weil's Understanding of Beauty and Justice," in *Attention: The Life and Legacy of Simone Weil*, E-journal, 1.2 (2021).
23 Vance G. Morgan, *Weaving the World: Simone Weil on Science, Mathematics, and Love* (Notre Dame: Notre Dame University Press, 2005), 3.
24 Christopher Bartlett, "Nautilus Spirals and the Meta-Golden Ratio Chi," *Nexus Network Journal* 21 (2019), 641–656. https://doi.org/10.1007/s00004-018-0419-3.
25 Morgan, 46.
26 Morgan, 3.
27 Simone Kotva, *Effort and Grace: On the Spiritual Exercise of Philosophy* (London: Bloomsbury, 2020), 176.
28 Rebecca J. Rose, "'I've Created a Monster!' On the Regrets of Inventors," *The Atlantic* (online: November 23, 2011).
29 Ibid.
30 Timothy Bella, "The Atomic Bombings Left Oppenheimer Shattered: 'I Have Blood on My Hands,'" *The Washington Post* (online: July 21, 2023).
31 Morgan, 2.
32 Murdoch, *Sovereignty*, 64.
33 Ibid., 73.

34 Eknath Easwaran, trans., *Bhagavad Gita* (Boston: Shambhala, 2004), 92.
35 Kotva, *Effort and Grace*, 150; *NB*, 94.
36 J. Ranilo B. Hermida, "Simone Weil: A Sense of God," *Logos: A Journal of Catholic Thought and Culture* 9.1 (2006), 133.
37 Benjamin P. Davis, *Simone Weil's Political Philosophy: Field Notes from the Margins* (Maryland: Rowman and Littlefield, 2023), 129.
38 Pétrement, *Simone Weil: A Life*, 62.
39 Emmanuel Gabellieri, "Reconstructing Platonism: The Trinitarian Metaxology of Simone Weil," in *The Christian Platonism of Simone Weil*, ed. E. Jane Doering and Eric O. Springsted (Notre Dame: Notre Dame Press, 2004), 145.
40 Ibid., 146.
41 Williams, "Necessary," 66.
42 Davis, *Simone Weil's Political Philosophy*, 25.
43 Rainer Maria Rilke, "Archaic Torso of Apollo," in *The Selected Poetry of Rainer Maria Rilke*, trans. S. Mitchell (New York: Random House, 1982), 61.

Part II

Plato and the Environment

6 Contemporary Dualist Ecological Readings of Plato's *Phaedrus*

In his investigation of nature and philosophy, John Sallis develops the historical conflict between Plato and the natural world:

> There is a certain Platonism that sets nature apart. It is not the Platonism of the dialogues, if indeed the disclosures achieved through the words, deeds, and myths of the dialogues can, in any coherent sense, be designated as Platonism. Rather, it is a Platonism that is set in position, that has undergone a transformation into a position that is placed in opposition to others.[1]

According to Sallis, the conflict or divide between human and nature is not inherent to Plato's dialogues, but rather has been read into them across the history of philosophy and then been used as either a convenient counterpoint or support for two polarized positions: human exceptionalism and human entanglement. Plato's *Phaedrus* is perhaps the dialogue that deals most with the theme of nature, and with this in mind, I will set out the dominant dualist reading of *Phaedrus* as explained by key figures in eco-feminism and ecological phenomenology. In these readings, Plato is interpreted as asserting an anti-materialist escape from the "lower" elements of a binary universe, namely, earth, woman, matter, and body, in order to pursue the "higher" pole of the Forms, the Good, and the masculine mind. In the next chapter, I will contrast this reading with Simone Weil's non-dualist reading of Plato, focusing especially on the role of *metaxu*. This Weilian non-dualistic interpretation will provide the foundation for the ecological ethics that I set out in the third part of the book. In agreement with Sallis, I believe the dualist interpretations of Plato come from the baggage of dogmatic "Platonists." As the thinkers this chapter engages will note, these dualistic interpretations have been taken up by Platonists across the history of western philosophy, but ultimately, I will argue that they are not the definitive interpretation to be taken from the dialogues themselves.

Phaedrus is a dialogue between the older Socrates and the young, beautiful Phaedrus, written by Plato in approximately 390 BCE, close to the time he is estimated to have written *Symposium* and *Republic*. *Phaedrus* focuses on the themes of love and rhetoric, and it is heavily laden with the theme of nature but often in an indirect way. These three themes overlap, inform, and interact creatively with

DOI: 10.4324/9781003449621-9

one another throughout the text, and they have been read by some as expressing the problematic hierarchical binaries found in Plato and his philosophical heirs in the western canon. The setting of the dialogue is unique in that it takes place beyond the Athenian city walls, "under a plane-tree, by the banks of the Illissus."[2] The choice to remove this dialogue from the domain of the city is unique for Plato, and the entirety of the dialogue plays upon a Socrates continually in the throes of paradox about his relation to nature versus culture, the sacred versus the profane, myth versus logic, and oral versus written tradition. Given the ecological concern of this book, I will focus on the theme of nature and mostly set aside the theme of rhetoric, but I begin with some reflections on the theme of love because it serves to illuminate both the concept of dualism and that of *metaxu*.

The dialogue opens with Phaedrus recounting Lysias' speech, which suggests it is best to have a lover who does not actually love his beloved. Lysias' reasoning for this is that love can pollute rationality and objectivity. As such, it is better to learn from a lover who is not in love with you. Socrates responds first on the same side as Lysias, but then unable to abide making a speech against love, he offers praise to the muses and to *Eros*. The obvious love between Socrates and Phaedrus offers a performative layer to the theme of love in that they cultivate and embody the type of love that they discuss. Indeed, Socrates claims that he knows Phaedrus as well as he knows himself, implying a level of loving attention and devotion to his interlocuter.[3] More than just orating the virtues and merits of love, Socrates enacts the idealized form of love between himself and Phaedrus, in which friendship and the life of the mind are prized over carnal acts of love. Rather than physically act on their love, Socrates chooses to remain under the plane-tree and talk with Phaedrus until the metaphorical and literal heat of the day has passed.

Diotima's ladder of love, which is shared as a part of Socrates' contribution to the speeches on love in *Symposium*, begins the journey towards idealized love of the Good with the beauty of a particular body arousing love and discourse between two people, moving up into a group of people, and continuing on towards the ideal form of love.[4] In this sense, love in the *Phaedrus* dialogue begins at the first rung of the ladder of love, as it is between two people instead of the multiple people often present for the dialogues. Socrates warns Phaedrus that in sexual love, one loses oneself to the material world and neglects the merits of higher spiritual pursuits. This is best described in *Phaedrus* by the allegory of the charioteer. Socrates describes the soul as having three parts: the charioteer who drives the horses; the white horse that strives for goodness; and the dark horse that strives for base pleasures. After beating the dark horse into submission in order to maintain a pure friendship with the beloved, the dark horse of the soul continues to desire physical intimacy, and the charioteer must resort to the "arguments of shame and reason" to suppress this intimacy.[5] The lovers' happiness and harmony hinges on this ability to enslave and beat the dark steed into submission.

Socrates claims that "the mind of the philosopher alone has wings" because the mind is the aspect of the human that reaches for the divine and leaves behind the vulgar earth.[6] To witness the beauty flowing from the eyes of her beloved makes the philosopher's wings begin to itch and ache in a deeply erotic metaphor. This longing is ruined by allowing pure love to be made base through the desires of the dark steed. The allegory of the charioteer and horses suggests a rigid dualism between matter and spirit, whereas the ladder of love in *Symposium* seems to suggest a more progressive form of love that shifts in direction but is not fully different in kind like two different horses. As such, *Phaedrus* can sometimes be used as a foil for materialist ecological thinkers and it is the key text to grapple with when considering alternate, non-dualist readings of Plato. The discussion about love itself is actually shorter than the discussion about rhetoric in *Phaedrus*, but the theme of love continues performatively throughout the remainder of the dialogue.

The theme of nature is present throughout this dialogue and continually inserts itself into the action of the speeches. Perhaps most notable is the anti-nature sentiment espoused by Socrates throughout the dialogue: "I am a lover of knowledge, and the men who dwell in the city are my teachers, and not the trees or the country," he tells Phaedrus by way of explanation for his seeming out of place as they walk along the river.[7] And yet, it is Socrates who corrects Phaedrus about the exact location on their walk where "Boreas is said to have carried off Orithyia," causing the reader to wonder how out of place and out of sorts Socrates truly is in the mythical natural environment.[8] Socrates also notes that nature leads one to myth, which leads to the sacred, but he follows this up with the suggestion that myth is a "crude philosophy" in comparison to self-knowledge.[9] Again and again, Socrates turns to nature and to myth in this dialogue as inspirations for inquiry, and his arguments about dialecticians and truth continually return to praising the sacred, the gods, and the muses as the benefactors of truth. He notes that being out in "holy"[10] nature causes him to be drawn into "a divine fury, for already I am getting into dithyrambics"[11] and as such he desires to go back into the city.[12] The dithyramb, a hymn sung by a chorus in honor of the god Dionysius, was commonly performed at festivals by a chorus of men or boys in Socrates' time and was known for decadence, improvization, and intoxication. The dithyramb is distinct from the measured and somber civilized paean hymn to the god Apollo. Here, Socrates seems to be rather afraid of falling into this Dionysian mode of poesis and wants to return to the controlled reason of the Apollonian dialectic and the city. But later, when he delivers his discourse in defense of love, he claims that divine madness is the gift of love and as such, it is a blessing.[13]

In supplying a reason for continuing his discourse with Phaedrus after they have concluded their discussion of love, Plato does not invoke the merits of dialogue or self-knowledge, but rather the presence of the cicadas in the plane-tree:

> And I believe that the grasshoppers chirruping after their manner in the heat of the sun over our heads are talking to one another and looking down at us. What would they say if they saw that we, like the many, are not conversing, but slumbering at midday, lulled by their voices, too indolent to think?[14]

From an ecological perspective, it is relevant that here we have a translation that uses "grasshoppers" instead of "cicadas." While it is unclear why this decision is made, it is interesting in that this points to a lack of specificity when it comes to the differences between cicadas and grasshoppers. If the grasshopper or cicada is simply a metaphor for something distinctly human, it does not particularly matter, but ecologically and biologically it matters deeply. I include this inconsistency in language because it is an interesting problem of translation and of merging philosophy and environmental science. But this fascinating discrepancy aside, Socrates recounts a myth in which cicadas act as a go-between, connecting the words of humans and the blessings of the muses. In this myth, the cicadas are the gatekeepers for philosophers to access inspiration: "They [the cicadas] win the love of Calliope the eldest Muse and of Urania who is next to her, for the philosophers, of whose music the grasshoppers make report to them … For many reasons, then, we ought always to talk and not to sleep at mid-day."[15] The cicadas have a special attention that is overlooked by humans and a special relationship with divine inspiration. They are closer to both nature *and* to the divine. This tale seems to be a reminder of the importance of the physical form and the balance between the material and the ideal. These liminal beings show that nature is not diametrically opposed to the divine in Plato's discourse but rather, that love of nature and inspiration of myth can act as connections to divine inspiration. The relationship between that which is natural and that which is supernatural is one of interconnected and interdependent ecosystems.

To the extent that we can count the character of Socrates in this dialogue as the vehicle for voicing Plato's own perspectives, it seems that Plato's relations to nature are problematic and even paradoxical.[16] I would like to note three final points about Socrates' and Plato's *supposed* preference for the urban and logical over the natural and mythic. The first two are in regard to Socrates. While his dialogues mostly take place within the city limits with masculine interlocutors using reason, his learning, particularly the learning he undergoes with Diotima, involves a woman known for divine intuitions and connected to a foreign land about a day-and-a-half's walk from Athens. This "visitor from Mantinea"[17] is known to have "procured for the Athenians, after they had performed sacrifices, a ten-year postponement" of the plague.[18] In a sense, she represents everything that Socrates is not and yet, she appears to be the one who teaches him the Socratic method: "I think it will be easiest to proceed as did my visitor from Mantinea with me on that occasion, by question and answer."[19] Diotima embodies mystical, poetic knowing. It seems that while Socrates fine-tunes his powers of discourse in the city, he learns how to learn in nature.

Secondly, Socrates famously "went about always barefoot"[20] and thus is already completely at home outside the city cooling his feet in the river, whereas Phaedrus simply happens to not be wearing his sandals on that day. Ferrari notes:

> Phaedrus, recall, left his sandals off by chance; but Socrates is always barefoot, and always "at leisure" – a leisure that is his life's work. Because as a philosopher the whole of life is Socrates' artistic concern, and not just a professionalised piece of it, his response to the environment is presented as spontaneous, welling directly from ordinary life rather than set against it in the manner of professional analysis.[21]

Socrates is at leisure, specifically grounded and in touch with whatever is beneath his feet, whether grass or cobbled streets. While most Athenians walk about the city in sandals, Socrates walks about barefoot, as if in the countryside. In *Symposium*, the spirit *Eros* is similarly "without shoes."[22] This is a sign of Love's poverty and yearning. Socrates' refusal of shoes represents simplicity and frugality, and it suggests an allegiance to the leisure of the countryside. It complicates his supposed loyalty to the city alone and allies him with the natural world and the nakedness of love. This reveals his primary mode as a lover of wisdom. Anne Carson reflects upon this phenomenon:

> The beginning has as its purpose to set us on the road
> that leads to the end.
> It directs our attention
>
> to our feet
> and asks us to remember
> the art of stepping along.
>
> The *Symposium*, for example, begins with Apollodoros
> going up the road to town (172a).
> He is stopped by a voice from behind
>
> and the story of the *Symposium* unfolds.[23]

For Carson, the fact that *Symposium* begins with the observation that Socrates unusually and notably is wearing shoes signals the reversal of norms that characterizes the text. It heralds a series of erotic inversions and shifting desires between the men. It signals that Socrates in particular lives in a liminal space between "drunk and sober, mortal and immortal, well shod or barefoot,"[24] and perhaps also between city and nature. Carson playfully ends her poem by noting that Diotima's feet remain a mystery,[25] nodding to the fact that at the heart of the dialogue is a mystical woman who offers up the truths of love as *metaxu*.

Thirdly, Plato's Academy, where he teaches philosophy years after Socrates' death and undoubtedly pens many of his famous dialogues, is located in a place not unlike the setting of *Phaedrus*: outside the city walls, along the river, in an area that was originally a sacred grove and later contained a religious precinct.[26] Rather than an appreciation of nature, this could be read as the destruction of nature, the suburban sprawl of the Academy as covering over a sacred grove in the same way that Platonist epistemology could be seen as covering over the worth of nature. Perhaps the truth is somewhere between these two extremes, but it certainly seems that both Socrates and Plato found sacred natural places to be conducive to thinking and learning, ideal sites for lovers of wisdom to meet and learn. Having set out these initial observations of some of the paradoxes about nature in the *Phaedrus* dialogue, we will turn to select eco-feminist and ecological phenomenologist interpretations of *Phaedrus* and the interpretation of Plato across the history of western philosophy. As the following thinkers will argue, when taken up in the history of western philosophy, the themes of *Phaedrus* are implicated in co-creating a hierarchy and a dualism in western thought that has been incredibly harmful to the west's relationship with the natural world.

Philosophical Responses and Readings of *Phaedrus*

The following thinkers serve to establish a particular dualistic interpretation of Plato that has been dominant in the western tradition. These thinkers do not subscribe to that tradition, but rather act as critics, calling out the issues that arise in a philosophy founded on dualism between spirit and matter, mind and body, man and woman, white and black, good and evil, and the realm of the Forms/the Good and the realm of the earth/base matter. Understandably, such an interpretation of Plato has led to a call by many contemporary thinkers, including those engaged here, to overturn the dominance of Plato in the western philosophical canon in light of the ethical harm caused to those relegated to the "lower" binary. Before we seek to establish an alternative reading of Plato, we will attempt to understand this binary interpretation of Plato through the criticism of those contemporary philosophers who suggest we should overthrow Platonism. This critique has come from myriad thinkers and philosophers, but I will call upon two eco-feminists and an ecological phenomenologist to explore the nuances of this critique. Carolyn Merchant offers an understanding of the history of this philosophical development of Plato by the Platonists, Val Plumwood explains why such a philosophy is problematic and an exclusionary ethical system, and David Abram brings a slightly more sympathetic voice to Plato but ultimately agrees that Plato's philosophy maintains a deeply problematic dualism.

Eco-feminism and the History of Platonic Thought

The term "ecological feminism" came from the French feminist Françoise d'Eaubonne circa 1974, who wanted to note women's unique potential to bring about ecological change. In this initial use of the word, "eco-feminism" was interested in myriad interdisciplinary ways that women could be understood as connected to nature. It was not until the 1990s that eco-feminism emerged as distinctly philosophical. Karen Warren explains that there are three general helpful characterizations of eco-feminism as a philosophical pursuit:

> it: (1) explores the nature of the connections between the unjustified dominations of women and nature; (2) critiques male-biased Western canonical philosophical views (assumptions, concepts, claims, distinctions, positions, theories) about women and nature; and (3) creates alternatives and solutions to such male-biased views.[27]

While this division begins as a gendered understanding of domination and denigration, it expands into what Plumwood will refer to as an "ethics of exclusion," which recognizes the assertion of a master subject over the natural world. This master subject is indeed master by way of gender, but also by way of race, class, ability, etc.[28] This expansion is key to a productive use of eco-feminism, as it allows for the nuances of intersectionality to play out and for gender to be fluid within this division. This is particularly relevant in coming to terms with the Platonist domination of nature and denigration of women along with other marginalized groups. The common reading of Plato that has formed the foundation for a hierarchical anti-earth, anti-woman, and racist discourse is well documented by eco-feminist thinkers and traces Platonist interpretation across the history of philosophy. Merchant's narrative of Platonic dualism in *The Death of Nature: Women, Ecology and the Scientific Revolution*, and Plumwood's critique of Plato in *Feminism and the Mastery of Nature* exemplify how Plato has been taken up in the eco-feminist sphere. While eco-feminists rightly point to how Plato has been taken up by the Platonists, to return again to Sallis, I will ultimately argue that the development of these ideas across the history of philosophy is not representative of Plato's potential, nor do I find the dominant Platonist reading to be a terribly convincing interpretation. This is not to say that I disagree with eco-feminism in general or Merchant and Plumwood in particular. Rather, it is to note that they are engaging with how the tradition has developed across history (the Platonists), whereas my predominate interest will be in returning to Plato's texts in themselves and suggesting that a relational reading of these texts is more aligned with the intention and spirit of Plato's work according to Simone Weil. Rather than disputing Merchant and Plumwood, I see myself as building on their work, albeit in a different direction. I am only free to return to the source and reimagine alongside Weil because they have established this predominant reading of Plato in light of its patriarchal ramifications and as such cleared the space for this creative reimagining. The first step in my process of reimagining Plato alongside Weil is this recognition of the patriarchal anti-nature reading of Platonic thought so that I can recognize the stakes and real-world consequences of this particular interpretation of Plato as it has been set out and examined by eco-feminist thinkers.

In *The Death of Nature*, Merchant traces the intellectual history of the link between woman and nature and the denigration of both.[29] Merchant notes that in Plato's *Timaeus*, the feminine world soul is placed in opposition to the ultimate unity of the masculine originary Oneness. To place this division in another context, we can consider a similar opposition in the Judeo-Christian Bible: "The earth was without form, and void; and darkness was on the face of the deep. And the Spirit of God was hovering over the face of the waters" (Genesis 1:2).[30] Here, the spirit of God is masculine, while the face of the deep, the void, the earth, is feminine. Plumwood importantly notes that the Platonic world soul is feminine only linguistically and not literally and as such, we cannot yet lay claim to a "men" versus "women" hierarchical dynamic, but remain at the linguistic divide between "masculine" and "feminine." The division that begins here will intensify across time. Following this initial division between the masculine spirit of God or absolute fullness and the feminine spirit of the world or absolute void, Merchant explains that Plotinus, the founding figure of what came to be called Neoplatonism, further divides the notion of the "lower" feminine soul into that which creates human souls from divine ideas and that which creates nature and the phenomenal world from matter. Thus, there is an initial rending of God and creation (in Weilian terms Creator and creature), then of creation into ideal soul (that which desires the Good) and material body (that which desires carnal pleasures). The material bodies of the natural world are thus twice removed from God according to Merchant.

Merchant notes that in western culture, the popular metaphor of earth as mother and nurse is drawn directly from Plato's *Timaeus*, it re-emerges in the Renaissance, and remains popular to this day.[31] There are myriad examples of this metaphor across various cultures and not all of them derive from Plato, but the particular dominant western canonical strain of thought that Merchant is following does point to this link between Plato and Christianity. The development of this binary division between masculine Good (God), feminine matter (the earth and also women), and a masculine embodied soul (men) moving from matter towards the Good, is increasingly entrenched into the genders, reaching its peak in the seventeenth century, with thinkers like Francis Bacon, who

> frequently described matter in female imagery, as a "common harlot." "Matter is not devoid of an appetite and inclination to dissolve the world and fall back into the old Chaos." It therefore must be "restrained and kept in order by the prevailing concord of things." [32]

At this point, the denigration of woman and nature is solidified in a way that is recognizable to contemporary western culture. The vulgar Platonic feminine nature becomes the imagery through which the great thinkers of the Renaissance conceptualize tragedy and suffering in the world.

Merchant adds to this the further denigration of woman and nature during the Renaissance, by thinkers like Marin Mersenne, who argues against *Timaeus*' feminine world soul in favor of a mechanistic earth: "he wrote that the earth did not have animal intelligence nor animal instinct: the world soul was neither the formal nor efficient cause of all earthly beings."[33] In this way, the living feminine earth is completely covered over, objectivized, de-animated, presented as if dead. The inert feminine earth remains the lesser aspect of the hierarchy and furthermore becomes object to the masculine subject, background to the action of the human intellect. Devoid of rationality, emotion, vitality and sentience, this mechanistic reading of woman and earth is read into the ontology of Plato that divides the masculine realm of the Good and the Forms from the feminine material earth. This reading covers over the problematic nature of this denigration. Afterall, one cannot hurt that which feels no pain, and thus man can do what he will to the mechanistic earth.

Merchant's history of the development of Platonist thought in relation to earth and woman explains the development of a hierarchical dualistic reading of Plato. These readings prove themselves to be incredibly damaging to anything or anyone that is given the "lower" tier of the hierarchical dichotomy. Particularly damaging is the work of cultivated ignorance implied in the mechanization of the lower dichotomy. Man does not even realize he is causing pain and destruction if he thinks that nature is a mere object. This makes for a mindset that perpetrates a "justified" violence against the lower side of the dichotomy because it understands that side to be there as an object "for" the use of the "higher" pole. For example, the natural world is an object present for the consumption, energy, and wealth of man, the wife is there for the service of the husband, etc. This leads to dire consequences for ecology as well as for human equity in regard to gender, race, ability, and social standing.

Plumwood on Plato and the Ethics of Exclusion

Plumwood turns to the duality of western thought and notes that women have traditionally been placed on the side of nature, necessity, and reproduction while men have claimed or been placed into the side of culture, freedom, and the creation of ideas:

> In terms of the assumptions of nature/culture dualism, women's "uncontrollable" bodies make them part of the sphere of nature. Such an assumption of women's "closeness to nature," where nature is taken as the realm of necessity over that of freedom, is of course extremely problematic for feminists.[34]

Similar to Merchant's take on Plato, Plumwood notes that "the body and the passions belong in Plato's account to a sharply distinct lower realm, homogenised and defined by exclusion, to be dominated and controlled by superior reason, and to be used in its service."[35] It is through this Platonic dualism, refined by John Locke during the Enlightenment, that nature is instrumentalized and people are made to think of the natural world as something that does not require or deserve moral consideration.[36] Picking up where we stopped with Merchant, in the Enlightenment-era western world, Locke suggests that nature is only imbued with worth by the imposition of man's labor:

> Though the earth, and all inferior creatures, be common to all men, yet every man has a property in his own person: this no body has any right to but himself. The labour of his body, and the work of his hands, we may say, are properly his. Whatsoever then he removes out of the state that nature hath provided, and left it in, he hath mixed his labour with, and joined to it something that is his own, and thereby makes it his property.[37]

Locke here notes the ownership each man has over his own body's labor and explains how this extends to the land when a man mixes his labor with that land. Locke then goes on to state that

> the grass my horse has bit; the turfs my servant has cut; and the ore I have digged in any place, where I have a right to them in common with others, become my property, without the assignation or consent of any body. The labour that was mine, removing them out of that common state they were in, hath fixed my property in them.[38]

In this section, Locke smoothly extends the ownership of his body's labor not just to the land that he works but also to the land worked by his animals and servants. The representation of a servant's work as the extension of the master's property has particularly troubling racial implications given the mass industry of slavery and Locke's own involvement in that industry.[39] As Carole Pateman and Charles Mills have argued, the social contract and property ownership in Locke come at the expense of marginalized others, whom Locke excludes based on what he sees as the incapacity or unwillingness to take up the responsibilities divinely ordained for human beings. These outsiders, namely women and racialized people in the arguments of Pateman and Mills respectively, become not signatories of the social contract, but rather objects that are a part of that contract.[40] Locke's contribution to man's enlightenment is at the expense of nature, racialized people, and women. Perhaps it is unfair to place Locke at the center of this debate, as Enlightenment thinkers in general had a penchant for the division of man and nature and for speaking loudly for the rights of some while ignoring the rights of others. But Locke is an apt thinker to take on this position, as he has been incredibly influential in the western canon and his work has been consistently upheld as exemplary in regard to empiricism, property rights, and political thought. With this in mind, he is well situated to also stand in for the racialized and gendered exclusions in social contracts if we keep in mind that he is but one player in a much larger social political philosophical narrative. Plumwood argues that the dualism Locke utilizes is born all the way back in Plato's ancient Greece. For Plumwood, western thought's link between reason and domination is the fatal flaw of our contemporary condition of ecological disaster and human alienation:

> The key to existential homelessness and to our denial of our dependence on nature is the dualistic treatment of the human/nature relationship, the view of the essentially or authentically human part of the self, and in that sense the human realm proper, as at best accidentally connected to nature, and at worst in opposition to it.[41]

Plumwood notes that the Platonic dialectic is not a simple "man versus woman" scheme, but instead, she sees it as an *ethics of exclusion.* The master subject is masculine but is also master by way of race, class, gender, ability, etc. In this way, Plumwood's feminist critique of Plato is intersectional in its recognition that there are multiple ways in which the Platonic subject identifies itself through exclusion:

> it is not only a masculine identity as such which underlies the Platonic conception of reason and of the life of reason, but a master identity defined in terms of multiple exclusions, and in terms of domination not only of the feminine but also of the slave (which usually combines race, class, and gender oppression), of the animal and of the natural world.[42]

This intersectional critique of Plato argues that it is not coincidental that Plato's writing is occasionally misogynistic but rather sees the Platonic denigration of women as fundamental to the entire system of Plato's thought. Women are necessarily represented as a series of lacks in order that man can be represented as a full subject worthy of striving to become a spiritual political being:

> Such consistent denial and derogation is no mere "casual misogynism." It is a major and extremely influential aspect of Plato's philosophical framework, involving the persistent association of women and their lives with the lower order of being, and the treatment of reason and of philosophy as the life of reason as oppositional to and exclusive of this lower order.[43]

Plumwood goes on to note Plato's supposed "feminism" in *Republic* based upon the possibility of women acting as Guardians and taking part in military exercises. For example, in his popular translation, Francis Cornford entitles this section of *Republic* "The Equality of Women," while Ernest Barker entitles this section of his translation "The Emancipation of Women." James Adams argues that Plato is attempting to better the position of women in Athens in this section of *Republic*, and Moses Hadas claims that women are treated equally to men in *Republic*. Conversely, Sarah Pomeroy argues that in fact, "Plato did not intend that woman be equal in status to men in the Republic."[44] Like Pomeroy, Plumwood rejects the suggestion of a Platonic feminism and instead suggests this as a tokenism in which a few elite women have the privilege and disposition to take on the "role of men." Plato's concern for a few elite women is not concern for all women, and these fully realized elite women hold their relatively high position in society due to the continued subjugation of others. The ethics of exclusion is maintained in this "feminism."[45] This exclusionary dualism is referred to by Plumwood as the "fault-line" of Platonic dualism:

> Platonic philosophy is organized around the hierarchical dualism of the sphere of reason over the sphere of nature, creating a fault-line which runs through virtually every topic discussed: love, beauty, knowledge, art, education, ontology. There are two sorts of being, two sorts of love, two sorts of equality, two sorts of knowledge, two sorts of causation, two sorts of art and even two sorts of music. In each of these cases, the lower side is that associated with nature, the body, and the realm of becoming, as well as of the feminine, and the higher with the realm of reason. The timeless, abstract realm of the Forms is separate and maximally distanced from the inferior "world of changes," of coming into being and passing away, and its representatives are everywhere treated as the superiors and masters of the representatives of the lower order.[46]

The distinction between the Platonic ideal realm of Forms and the earthly realm of illusions implies the necessary exclusion of certain subjects and beings.

Even if we generously read Plato as offering a universe that marries reason and matter, Plumwood argues that the marriage is one of patriarchal domination.[47] She goes on to argue that the world soul is not an enchantment of nature, but rather, nature's colonization.[48] The predominant metaphor used by Plato to conceptualize this marriage is that of master and slave.[49] This is best exemplified in the charioteer analogy of the soul in *Phaedrus*.[50] Plumwood notes that the depiction of the charioteer with his white and black horses is particularly racialized and enforces a master/slave relationship between these aspects of the soul. However, she recognizes that unlike the earlier dualism of *Phaedo*, which denotes the body as separate, externally fettered to the material, and distinct from the soul, *Phaedrus* (as well as *Republic*) internalizes the dualism. The soul contains both material and spiritual desires, and it furthermore contains the driver directing and taming those desires.[51] This distinction is pivotal to non-dualist readings of Plato and while it helps make sense of the reality of our internal struggles and the experience of being split against ourselves, Plumwood argues that it creates the very idea that the body and soul should be thought of as distinct in the first place. For Plumwood, Plato creates the very problem of dualism that he sets out to overcome.[52] In positing this dualism as the originary state of our ontological condition, it becomes nearly impossible to remedy this rift. The strongest solution to this problem would be not to begin with the supposition of division in the first place.

Plumwood notes that while in our post-Nietzschean world in which the "death of God" has been widely agreed upon (and it is no longer intellectually common to believe in something like the realm of the Forms or heaven), we have maintained the structure of elevating ourselves as humans over and against nature.[53] Indeed, while many of the foundational ontological notions of Plato (the Good, the realm of the Forms, the shadowy existence of the Cave) have been critiqued or dismissed in the wake of thinkers such as Friedrich Nietzsche, Karl Marx, and Sigmund Freud, allegiance to many of the epistemic Platonic outcomes remains. This includes what Plumwood refers to as a reversal of common sense: "The Platonic theory appears to represent an extraordinary inversion of intuitive thinking: the lifeless world of the Forms gives eternal life, the living world of nature is called a tomb."[54] Plumwood argues that the Platonic inversion of practical thinking remains prevalent in the contemporary world. This inversion enforces the ethics of exclusion with particularly dire consequences for those marginalized by being placed within the realm of the lower hierarchical divide, including the ecology. From the eco-feminist perspective, we see how the Plato of inversion and invisibility acts as a force of oppression. Furthermore, we see Plato as the father of the dualism between matter and ideal, and as such the harbinger of a particular hierarchical, patriarchal structure of thinking.

From Oral to Written Word and from Community to Individual

The final text that I will call upon to clarify the contemporary ecological dualist reading of Plato is David Abram's eco-phenomenology text, *The Spell of the Sensuous: Perception and Language in a More-Than-Human World.* In this book, Abram draws together the phenomenological method of studying the structures of experience, or consciousness, as they help us to create meaning and belonging in the world and applies it to our experiences of the larger ecology and to philosophical environmental questions. Originating with the philosopher Edmund Husserl

(1859–1938), phenomenology has been a major influence in continental philosophy since its inception. Drawing together the phenomenological method with environmental thought is, coincidentally, precisely what this book seeks to do with Weil's own method, which is not properly in the school of phenomenology, but as Mark Shiffman notes, "Weil seems to be working in the direction of a phenomenology, but one that does justice to the mind's experience of otherness rather than methodologically bracketing being."[55] Indeed, Lissa McCullough argues that "Weil was a pioneering phenomenologist of the body, for remarkably like [Maurice] Merleau-Ponty – but beginning a decade and a half before him in 1930 – Simone Weil's thinking centered on the foundational or axial role of the body in structuring thought and ordering the world."[56] But even if we follow McCullough in her suggestion that Weil is a phenomenologist akin to Merleau-Ponty (which I do), Abram's interpretation of *Phaedrus* will in the end align with eco-feminism rather than with my proposed Weilian reading because ultimately, Weil was incredibly unique among her phenomenologist/existentialist peers in her interpretation of Plato. Thus, from Abram, we can further layer and fill out the dualist interpretation of Plato (and the Platonists) set out by Merchant and Plumwood.

Abram notes that Plato "carefully developed and brought to term the collective thought-structures appropriate to the new technology" of reading and writing, which was sweeping through Greek culture at that time.[57] In Abram's reading, Socrates' insistence that his interlocutors clarify and define their arguments acts to separate the individual and her thinking process from the mythical narrative repetition which acted as the mode of knowing and thinking in oral cultures until that point. It is Socrates' dialectic that works to separate the "I" from the world, from myth, and from history and in so doing rends that which had been inseparable.[58] In Socrates' questioning, the "I" separates itself from the unreflective immersion in myth. Socrates dialogues and thinks only with spoken words (not with writing) and as such establishes himself as the father and origin of this tradition of dialectic engagement that has shaped much of western institutional philosophy. Similar to Plumwood when she states that Plato creates the very mind–body dualism that he seeks to overcome, Abram sees the Socratic dialectic as fashioning the I that is distinct from the surrounding world, therein producing the division that can only be overcome through the epistemic return to the Good by acquiring greater knowledge in beauty and love. Socrates creates the problem he seeks to solve.

According to Abram, the shift that happens as society moves from oral to written culture is what Socrates is grappling with in *Phaedrus*' discussion of rhetoric. Abram argues that "all discourse, even written discourse such as this, is implicitly sensorial and bodily, and hence remains bound, like the sensing body, to a world that is never exclusively human."[59] Abram notes the body's participation in language. In both written and spoken words, he notes the interconnection between language, the world, and other beings. But writing also allows for an individual human's separation from the world. For Abram, it is writing that gives the possibility for positing the "unchanging, visible form independent of the speaker."[60] This constitutes the separate Platonic realm of the Forms, which is often understood as an ideal realm that exists in opposition to the material earth:

> For the letters of the alphabet, like the Platonic Ideas, do not exist in the world of ordinary vision. The letters, and the written words that they present, are not subject to the flux of growth and decay, to the perturbations and cyclical changes common to other visible things; they seem to hover, as it were, in another, strangely timeless dimension.[61]

In addition to his comparison between writing and the Forms, Abram further notes that writing as a technology shapes the way in which humans create themselves and distinguish themselves over and against the world: "we may be sure that the shapes of our consciousness are shifting in tandem with the technologies that engage our senses – much as we can now begin to discern, in retrospect, how the distinctive shape of Western philosophy was born of the meeting between the human senses and the alphabet in ancient Greece."[62] In this way, he makes a compelling argument for the birth of western philosophy as the creation of both the individuated subject and the separation of one's self from the surrounding world with its larger ecosystem of existence.

Socrates and the Voice of Nature

Abram draws attention to Socrates' disdain for the natural world in the *Phaedrus* dialogue and he compares this to the Homeric world from which Socrates is seeking to break free, with its close ties to the gods and the natural world. While Abram does not draw out all of the Socratic irony and the paradoxes in Socrates' *alleged* disdain for nature, he does note that despite the ambivalence to writing and to nature, the dialogue takes place in the written form and is set in nature: "Unlike the other Platonic dialogues, the *Phaedrus* alone occurs outside the walls of the city, out beyond the laws and formalities that enclose and isolate the human community from the more-than-human earth."[63] Abram argues that in a sense, the two men embark on a vision quest, testing the limits of philosophy "by opening and exposing it to the non-human powers that for so long had compelled the awe and attention of humankind."[64] For Abram, this is in contrast to Plato's *Republic* and its vilifying of the old *natural* gods: "Plato brings philosophy itself outside the city, there to confront and come to terms with the older, oral ways of knowing which, although they may be banished from the city, nevertheless still dwell in the surrounding countryside."[65] It is from within this rural realm of knowing that Socrates can critique the written way of knowing and express respect for the "oral, animalistic universe that is on the wane."[66] For Abram, the city is bound to writing and nature is bound to orality.

In his tale of the cicadas, Socrates takes on the characteristics of Indigenous, oral storytelling cultures, according to Abram.[67] Socrates' story states that the cicadas were once humans who sang and rejoiced all day, forgetting to eat. The material is ignored in favor of the ideal. They are then reincarnated as cicadas and as such, all they can do is sing without regard for their fleeting physical form. Giovanni Ferrari notes that it is an excellent example of *contrapasso*, or Dante's suggestion in the *Divine Comedy* that the punishment fit the crime.[68]

Abram points to the reverence that Socrates shows towards the cicadas, which indicates "although not without a sense of irony, the respect that is properly due to such insects, who might confer a boon upon the two of them in return."[69] The admiration shown here reveals an attempt to reconcile the natural mystical world with Socrates' ideal urban cultural world, according to Abram. Ferrari goes so far as to argue that in sharing the myth of the cicadas, Socrates suggests that it is nature rather than culture that listens and hears the Good and notices human praises of the divine. Nature is our avenue to communication with the Good and access to the realm of the Forms.[70] But Abram reads this reconciliation with nature as a simultaneous subtle devaluation of nature. In Socrates' tale of *Eros*, the bodily desires and the sensory world are to be honored but only in so far as they lead one beyond the earth to the eternal Forms.[71] Abram's is a tempered reading of Socrates as more ambivalent than reverent towards the cicadas. Ultimately, while Abram offers a sympathetic reading of *Phaedrus*, he argues that Plato enforces a hierarchical idealism. For Abram, even within the most generous of Plato's texts, Plato still subtly degrades the material earth and oral nature-oriented ways of knowing while uplifting the ideal realm of the Forms cultivated in the city. Abram offers a generous reading of Plato, and *Phaedrus* in particular, that nevertheless maintains the interpretation of Plato as a foil for ecological materialists. Plato acts as the ultimate idealist, anthropocentrist, human exceptionalist, dualist, and as such, an apt antithesis to major tenets of eco-feminism and ecological phenomenology. Like Merchant and Plumwood, Abram ultimately reads Plato as the source of an ontology that denigrates mystery, disregards ecology, perpetrates an ethics of exclusion, and reveals the very foundation of the western philosophical tradition as patriarchal to its very core. This establishes, therefore, everything that we might hope to gain through an alternative reading of Plato, namely difference, diversity, ecological justice, social justice, ethics of inclusion, and relationality. In the following chapter I will set out the non-dualist reading of Plato by Simone Weil in order to explore a Weilian ecological ethics.

Notes

1 John Sallis, *The Return of Nature: On the Beyond of Sense* (Indianapolis: Indiana University Press, 2016), 60.
2 Plato, *Phaedrus*, trans. Benjamin Jowett (Bulgaria: Demetra Publishing, 1870), 4.
3 Ibid., 228a, p. 6.
4 Plato, *Symposium*, ed. M.C. Howatson and Frisbee C.C. Sheffield, trans. Howatson (Cambridge: Cambridge Press, 2008), 210a–212c, p. 48–50.
5 Plato, *Phaedrus*, 256a, p. 47.
6 Ibid., 249c, p. 37.
7 Ibid., 230d, p. 10.
8 Ibid., 229c, p. 8.
9 Ibid., 229e–230a, p. 9.
10 Ibid., 238d, p. 21.
11 Ibid.
12 Ibid., 242a, p. 26.

13 Ibid., 245c, p. 31.
14 Ibid., 259a–b, p. 51.
15 Ibid., 259d, p. 52.
16 It is not clear that Socrates' voice is necessarily a mirror of Plato's own thoughts, nor is it clear that Socrates' thought is adequately represented in the dialogues of Plato. One school of thought that recognizes a distinction between Platonism and a Socratic philosophy was spearheaded by Gregory Vlastos in the 1990s. See Gregory Vlastos, "Socrates Contra Socrates in Plato," in *Socrates: Ironist and Moral Philosopher* (Cambridge: Cambridge University Press, 1991), 45–80.
17 Plato, *Symposium*, 211d, p. 50.
18 Ibid., 201d, p. 37.
19 Ibid., 201e, p. 37.
20 Ibid., p. 3.
21 Giovanni R.F. Ferrari, *Listening to the Cicadas: A Study of Plato's* Phaedrus (Cambridge: Cambridge University Press, 1987), 21.
22 Plato, *Symposium*, 203c, p. 40.
23 For more, see Anne Carson, "Shoes: An Essay on How Plato's 'Symposium' Begins," *The Iowa Review* 25.2 (1995), 47–51. www.jstor.org/stable/20153663.
24 Ibid., 50.
25 Ibid., 51.
26 Rihll notes the Academy's remoteness from the Agora, but Trelawny-Cassity notes that the distance was not so far as to hinder any Athenians interested in going there. See T.E. Rihll, "Teaching and Learning in Classical Athens," *Greece and Rome* 50.2 (2003), 174. Lewis Trelawny-Cassity, "Plato: The Academy," in *The Internet Encyclopedia of Philosophy* (online: ISSN 2161-0002).
27 Karen J. Warren, "Second Kind of Position in Feminist Environmental Philosophy: Ecofeminist Philosophy," in "Feminist Environmental Philosophy," *The Stanford Encyclopedia of Philosophy*, Edward N. Zalta (ed.) (2015). https://plato.stanford.edu/entries/feminism-environmental/#SecKinPosFemEnvEco.
28 Val Plumwood, *Feminism and the Mastery of Nature* (New York: Routledge, 1993), 72.
29 I will refer the reader to Merchant's work for a full understanding of the historical movements involved in this process: Carolyn Merchant, *The Death of Nature: Women, Ecology, and the Scientific Revolution* (New York: Harper, 1990).
30 *The Holy Bible*, Genesis 1:2. See also Catherine Keller, *Face of the Deep: A Theology of Becoming* (New York: Routledge, 2003), XVIII.
31 Ibid., 27.
32 Merchant, 170.
33 Ibid., 196.
34 Plumwood, 37.
35 Ibid., 70.
36 Ibid., 70.
37 John Locke, "Property," in *Second Treatise of Government*, ed. Dave Gowan and Chuck Greif (online: Project Gutenberg, 2003), section 27.
38 Ibid., section 28.
39 It is contested among scholars of Locke whether his involvement in the Royal African Company and other Africa slave trades is separate from his philosophical work. For example, Hinshelwood argues that Locke's philosophy was directly affected by the enslavement he saw in the Carolinas, whereas Brewer argues that the presuppositions so often attached to Locke do not properly fit into his own work but rather into the political

laws and policies of the Stuart Kings that Locke opposed. See Brad Hinshelwood, "The Carolinian Context of John Locke's Theory of Slavery," *Political Theory*, 41.4 (2013), 562. Locke is also generally indicted not only for his financial investments in the slave trading companies but also for his role in writing the *Fundamental Constitutions of Carolina*, which included rights of slave owners. Some scholars say he was active in its authorship, others say he is merely guilty by association through his service to one of the Lord Proprietors of the Carolinas, Anthony Ashley Cooper. Also see Holly Brewer, "Slavery, Sovereignty, and 'Inheritable Blood': Reconsidering John Locke and the Origins of American Slavery," *The American Historical Review* 122.4 (2017), 1042–1045. In the particular section "On Property" that I reference here, Locke writes of servants not slaves. The argument for slavery involves ownership of a whole person and not just buying or renting their labor power. This is justified by imagining some people as congenitally incapable of or willfully resistant to obeying divine law, which Locke believes should be obvious to anyone whether or not they have been exposed to Christian teachings. The divine law in question gives humans the responsibility for developing natural resources which God provides for the purpose of development. Those who have proven themselves (through the "uncivilized" state of their culture) as unable or unwilling to do this do not have the rights God gave humans to be the sole owners of our labor. Thus, some persons are imagined in Lockean theory as natural resources who need to be developed rather than as persons with a right not to have their bodies controlled by another or their labor claimed as the private property of another (as in slavery).

40 See Carole Pateman, *The Sexual Contract* (Stanford: Stanford University Press, 2018); Charles W. Mills, *The Racial Contract* (Ithaca: Cornell University Press, 2022).

41 Plumwood, 71.

42 Ibid., 72.

43 Ibid., 77.

44 For Plato's discussion of the role of women in his Republic, see Plato, *Republic*, 454d–457d, p. 116–119. See Francis Cornford, *The Republic of Plato* (Oxford: Oxford University Press, 1965), 144–155; James Adam, *The Republic of Plato* (Cambridge: Cambridge University Press, 1963), 280; Moses Hadas, *A History of Greek Literature* (New York: Columbia University Press, 1953), 138; Ernest Barker, *The Political Thought of Plato and Aristotle* (New York: Franklin Classics, 1959), 144–145; Sarah Pomeroy, "Feminism in Book V of Plato's Republic," *Apeiron* 8.1 (1974), 33; Nancy Tuana, *Feminist Interpretations of Plato* (University Park: Pennsylvania State University Press, 1994); Susan Moller Okin, *Women in Western Political Thought* (Princeton: Princeton University Press, 2013).

45 Plumwood, 79.

46 Ibid., 81.

47 Ibid.

48 Ibid., 86.

49 Ibid., 87.

50 Ibid., 89.

51 Ibid., 90.

52 Ibid., 91.

53 Ibid., 101.

54 Ibid., 97.

55 Mark G. Shiffman, review of *Weaving the World: Simone Weil on Science, Mathematics, and Love*, by Vance G. Morgan, *Notre Dame Philosophical Reviews* (2006)

(online: https://ndpr.nd.edu/reviews/weaving-the-world-simone-weil-on-science-mathematics-and-love/).

56 Lissa McCullough, "Simone Weil's Phenomenology of the Body," *Comparative and Continental Philosophy* 4.2 (2012), 196.

57 David Abram, *The Spell of the Sensuous: Perception and Language in a More-Than-Human World* (New York: Vintage Books, 2017), 108–109.

58 Ibid., 109.

59 Ibid., 297.

60 Ibid., 112.

61 Ibid.

62 Ibid., 115.

63 Ibid., 117.

64 Ibid.

65 Ibid.

66 Ibid., 118.

67 Plato, *Phaedrus*, 259b–d, p. 52. Abram's authority to write about Indigenous ways of knowing and being has been challenged by some Indigenous thinkers. For more, see Lorraine Brundige and Douglas Rabb, "Phonicating Mother Earth: A Critique of David Abram's *The Spell of the Sensuous: Perception and Language in a More-Than-Human World*," *Ayaangwaamizin: The International Journal of Indigenous Philosophy* 1.2 (1997), 78–88.

68 Ferrari, 27.

69 Abram, 121.

70 Ferrari, 26.

71 Abram, 122.

7 A Non-dual Reading of Plato via *Metaxu* (μεταξύ)

Having established the dominant dualist reading of Plato in the previous chapter through an examination of Merchant, Plumwood, and Abram, I will now turn to the possibility of a non-dualistic reading of Plato, which might be realized, as Weil suggests, through Plato's concept of *metaxu* or *mediation*. After establishing the concept of *metaxu*, I will turn to various non-dualistic readings of Plato offered by Catherine Pickstock, Jennifer Rapp, and Giovanni Ferrari in order to showcase that while this is certainly less common in Plato scholarship, it is by no means a project unique to Simone Weil. Having situated Weil among these other non-dualistic readings, I further unpack Weil's interpretation of Plato to establish a strong argument for why her interpretation can and should be used to provide a novel form of ecological ethics.

The distinct but related movements of *metaxu* (the *in-between* or the *mediation*), transcendence (the ladder of love towards the Good), and immanence (the descent that brings the ethical Good into the world through love and beauty) are exemplified in Plato's Allegory of the Cave. In this allegory, prisoners are chained in darkness and mistake the shadows cast on the wall from a fire to be the truest reality of existence. One prisoner has the epiphany or the moment of grace that makes her turn away from the shadows. She glimpses something beautiful in these shadows, which delivers the experience of a connection, a mediation, or a bridge between her reality and something else. The *metaxu* prompts her to turn and to ascend from the cave into the light. Allegorically, she climbs up the ladder in an epistemically cultivated process of transcendence: she acquires greater knowledge and love of the beautiful as she transcends. But when she reaches the surface, she cannot stay above in the sun and instead, she is called by her ethical duty to descend and return to the cave for the others. Symbolically, this is the moment of ethics and immanence in which the light or the good is shared with others. So, there are three distinct movements at play: (1) the moment of *metaxu*, or the link between natural and supernatural, (2) the *transcendence* or the climb through epistemic growth, and (3) the *immanence* or the descent that shares the good with those still below. The *metaxu* is a material thing that points one towards the possibility of transcendence but is not itself that transcendence.

DOI: 10.4324/9781003449621-10

According to Gary Scott and William Welton, *metaxu* is the crux of a third way to read Plato's dialogues. They argue that there are two traditional ways of reading Plato: (1) the dogmatic or earnest reading that is attached to Plato and assumes real knowledge can be gained and asserted, and (2) the skeptical reading that is attached to Socratic ignorance or the recognition that our highest wisdom is in recognizing our lack of wisdom and as such, all we can learn from the dialogues is the impossibility of absolute truth or knowledge.[1] In the developmentalist reading of Plato, both of these views exist in Plato's writing at different stages. Plato begins with the skeptical Socratic philosophy but over time develops his own dogmatic philosophy: "From this point of view, philosophy as the comprehension of Forms is Plato's later conception of philosophy, one that provides a way out of the skepticism of the Socratic position."[2]

In the dogmatic reading of Plato, ethical political knowledge is provided by a Philosopher King with his ability to transcend to the Good and the Forms. Here, one reads Plato in earnest and as such interprets him as a thinker with a strict ethical political code founded on a dualistic view of a world divided between material and ideal realms. In this dogmatic interpretation, one can receive absolute knowledge from the Forms. This interpretation is a positive fulfillment of one's desire for wisdom and is aligned with Plato himself (as opposed to the wisdom of Socrates). This reading is precisely the interpretation of Plato that emphasizes a hierarchical dualism between matter and ideal. It is in this reading that an ethics of exclusion is foundational for transcendence into ideal knowledge. According to Scott and Welton, this reading emphasizes the Platonic Forms as the most important aspect of Plato's philosophy and in contemporary Plato scholarship this reading is referred to as "doctrinalist" in that it takes Plato to be asserting his own doctrines through the dialogues.[3]

The second reading, less popular but still ever-present in Plato scholarship, is the "non-doctrinalist" skeptical reading, which is founded upon Socratic ignorance. In the skeptical reading, to gain knowledge is only to recognize one's own ignorance through the process of dialogue. Here, knowledge is not fixed, but continually changes. Rather than reading Socrates as the mouthpiece for Plato's opinions, this interprets all of the opinions offered in the dialogues as equally valid. The most important part is staying open in a continual state of epistemic skepticism: "The skeptical view presents philosophy as the examined life, a Socratic quest that involves a developing awareness of one's own ignorance."[4]

Thus, the dogmatic interpretation of Plato offers a positive epistemology in which true knowledge can be attained through knowledge of the Forms, whereas the skeptical interpretation offers a negative epistemology, in which no knowledge can ever be truly attained but can be hinted at through dialogue. The developmentalist reading is one in which Plato's earlier texts are read as skeptical with an emphasis on Socratic irony and his later texts are read as dogmatic with an emphasis on the ideal Forms. Scott and Welton push back against the developmentalist reading, because they see *both* the Forms *and* Socratic wisdom as present in the dialogues. The Form of one thing or another, which implies a clear truth, is never explained as a clear absolute thing, but rather through the dialectic of a philosophical dialogue.

There is a gap between the medium and the message that leads them to a proposed third way of interpreting Plato:

> Were it possible for the philosopher to easily and securely grasp a Form and to express its perfection directly in language, it would seem that the dialogic and cryptic character of the dialogues would conflict with Plato's rationalist conception of philosophy; but if the Forms remain elusive, even as dialectic enables the philosopher partially and imperfectly to recollect them, then the dramatic and elusive character of the dialogues make an essential contribution.[5]

The third way that Scott and Welton propose combines non-propositional knowledge, or procedural knowledge that cannot be logically verified, with the Platonic "intermediary" or the *metaxu*. This third way suggests that "Plato conjoins the 'negative' experience of recognizing one's ignorance with the 'positive' experience of coming to desire the wisdom one lacks."[6] In this way, the Forms are considered alongside Socratic ignorance and not in opposition to it. This requires a weaker, less reified concept of the Forms:

> The Forms are never presented as dogmatically secure possessions, but rather as glimmering desiderata, the objects of a quest, objects that can inspire us, but which continually elude us in some way. Yet even in order to so elude us, they must also somehow be present to us, open to examination and inquiry.[7]

With the strict division of Platonic epistemology between the first sense of "negative" Socratic wisdom, followed by the second "positive" realm of the Forms, the intermediary is lost and the importance of our position in-between is discarded: "Plato's use of the notion of 'Forms' is misunderstood if it is seen outside the context of such intermediacy. Reflecting on the notion of intermediacy will serve to illuminate the simultaneous connection and tension in Plato's dialogues between their 'Socratic Ignorance' and 'Platonic Forms.'"[8] In this liminal sense of knowledge, we have a reading of the Forms as *temporal* rather than *spatial* in that they can appear to us in moments of knowing, but never completely nor absolutely and certainly not in a space physically beyond our own world. As such, the distinction between the material realm and the ideal realm of the Forms is overcome in a physical/spatial sense. The moments of *metaxu* connect not just these two distinct epistemologies (skepticism and positivism) but also connect the body–mind dualism that is implied in placing the realm of the Forms in space rather than in time.

As a larger philosophical movement, metaxological philosophy attempts to apply the *logos* to this "in-between." William Desmond and Kevin Kennedy define metaxological philosophy as follows:

> "*Metaxu*" is the Greek word for "between," while "logos" can mean an accounting, or reasoning, or wording. A metaxological philosophy is concerned with a logos of the *metaxu*, or a wording of the between. Such a philosophy is

> concerned with life itself as a between space, a *metaxu*, and with the fact that this between is an articulated middle or intermedium. A metaxological philosophy is concerned with wording the between.[9]

Bringing together *metaxu* and *logos* aligns with Scott and Welton's attempt to bring together non-propositional knowledge with the intermediary. Additionally, these ways of thinking are akin to Weil's assertion that we must not mistake bridges for destinations:

> The bridges of the Greeks. We have inherited them but we do not know how to use them. We thought they were intended to have houses built upon them. We have erected skyscrapers on them to which we ceaselessly add stories. We no longer know that they are bridges, things made so that we may pass along them, and that by passing along them we go toward God.
>
> (*NB*, 370)

The emphasis on intermediaries means that there is no final destination for our knowledge, no spatial realm of the Forms that we may access, but rather a crossing over, a becoming that is moving towards moments of the ultimate Form of the Good.

We should note, alongside Scott and Welton, that *metaxu* in Platonic thought involves various kinds of intermediaries: that of recollection, of right opinion, and even of philosophy as a whole.[10] *Metaxu* is a theme and an overall structure of Platonic thought; it is also used in diverse ways to describe that which is in-between (bridges or links via the beautiful) in tandem with that which transcends (ladders via epistemic desire) in order to descend into immanence (ethics via eros).

While in her own philosophy Weil will have a particular reading of *metaxu* as those things, places, or events that create a bridge between the natural earthly realm and the supernatural ideal realm of the Forms, she also notes the myriad ways that the theme of intermediaries plays out in Plato's thought. For example, Weil notes that in *Symposium*, the original people who are torn in two in Aristophanes' speech are a metaphor for our ontological status: "Our vocation is unity. Our affliction is to be in a state of duality" (*IC*, 11).[11] To be torn asunder from one's lover is not righted by finding a human partner but rather is a recognition of the human condition as being at once separate and unified. There is no cure for this condition but only moments of *metaxu*. It is not an external but an internal tear. Similarly, Weil claims that Agathon and Phaedrus are *both* correct in their contradictory claims that Love is young, and that Love is old. For Weil, Plato's notion of love must be both "absolutely ancient and absolutely young" (*IC*, 114). Love is the first cause, which begins creation and as such it is *a priori*, but it is also ignited anew in a unique way by particular things. And while force dominates the material world and holds sway over all that is created, love is that which we desire, and which leads humans to the possibility of grace. Necessity descends upon humans, but it is followed by the descent of love. Both are interwoven in the beauty of the world. Weil's is a metaxological reading of love as a link between an orientation towards Plato's realm of the Forms (the first, dogmatic reading) and the assertion of

Platonic ignorance (the second, skeptical reading).[12] The metaxological readers of Plato focus on bridge-building rather than dualism and earthly alienation.

Beyond Dualism

If western canonical philosophy is born of violent, hierarchical dualism and that is its sole legacy, then there is a temptation to reject the western tradition, which could also imply an abandonment of responsibility for the violence which that system has perpetrated. One potential hermeneutical method that would avoid this can be found in Catherine Keller's reinterpretations of biblical text. In her book *Face of the Deep: A Theology of Becoming*, Keller investigates the possibilities of the *ex nihilo* and the *deep* (*tehom*) from which creation occurs in order to imagine new possibilities in the Abrahamic traditions. Keller notes:

> *Face of the Deep* labors to reclaim for theological reflection the *tehom* inscribed "in the beginning" itself. I do not pretend to construct its opening as it ever was, but as it might yet be: as it unblocks possibilities within, at least, a particular postmillennial context. I read these possibilities as divine gifts and graces.[13]

Keller's work suggests that the dominant interpretation of a text has real effects in the world but that does not mean that it is the only, the definitive, or even the correct interpretation. This creative ethical mode of interpretation is precisely what I aim to bring to my reading of Plato.

In the case of the interpretation of Plato, a dualist separation of the material earth and our bodily existence from the realm of the Forms, ideas, and the divine Good has formed the foundation for much of western thought and Christianity. But that need not be the way we continue to read his work. Following Keller's style, I do not wish to assert *what was* or even *what is*, but *what is possible* in the western tradition. This approach may also let us stay with the trouble of environmental ethical issues in order to comprehend where we have gone wrong and how we can do better. This return to the Platonic origins of the canonical western philosophical tradition in order to reinterpret and reimagine the latent possibilities of that tradition calls upon readers to be creative and bold but also to consider that the inheritors of the western tradition should take responsibility and be held accountable for the violence and damage this tradition has so often perpetrated. For Weil, this responsibility means actions to protect other beings, as we shall see.

In her non-dualistic reading, Pickstock argues that Socratic orality in Plato is "an account of the subject as doxological."[14] By doxology, Pickstock is referring to a subject whose *raison d'être* is the praise of the divine. With this in mind, Pickstock asserts that Platonic ethical practice entails an orientation towards liturgical praise of the divine more than an ideal of rational contemplation. This is seen in *Phaedrus* with Socrates' recantation of his first speech against love and the subsequent prayer he offers in praise of Eros. Rather than rationality as his motivating foundational characteristic, Pickstock sees Socrates as driven by cultivating and embodying divine praise. Turning to one of the most damning metaphors in *Phaedrus*, so far as an implied hierarchical duality – that of the charioteer – Pickstock focuses on the Socratic soul. Rather than emphasizing the violent and external domination of gendered or racialized others that can be read into Socrates' description of the soul as charioteer guiding dark and light horses, Pickstock argues that this myth is meant to suggest an internal reconciliation necessary for cultivating a state of divine praise. This distinction between internal and external struggle is imperative for the possibility of ethical difference in Plato and as such, for an ecological ethics that holds open a between-space for the things, beings, humans around us to exist on their own terms. Pickstock explains:

> Socrates' declared intention in telling the myth is to reconcile erotic passion with the philosophic quest for the good, and at the close of the speech, he describes exemplary philosophic lovers, and the benefits to both lover and beloved reciprocally bestowed via the mediation of erotic beauty. He asserts that both lovers gain a good greater than that attained by either human moderation or divine madness (256b).[15]

Pickstock reads the metaphor of the charioteer as maintaining a balance *between* inward and outward orientations – between reaching towards the beloved and holding back with cautious, respectful restraint. She suggests that it is not a rejection of the rational discipline of the city and dialogue with interlocuters nor of the divine madness of nature and the sacred, but a third way that mediates or "bridges" the good and the beautiful, the ideal and the material.

Ferrari describes this reconciliation of *logos* (written word, "culture," individuals in dialogue) and *mythos* (oral culture, "nature," Homeric recitations) by arguing that rather than a progression from one into the other, myth is always in the background of how we logically know what we know. Myth is always one step ahead in that the stories of mythology form how we see the world that we walk in: myth explains the philosopher's prior intuition of that which she investigates as a recollection of the Forms.[16] Logic explains *what* is here in a rational way, while myth helps us to intuit *why* this matters and to get a sense of the origins behind what immediately presents itself. Myth and logos must work alongside one another to create a meaningful experience of the world. In this way, Ferrari sees the movement between mystery/myth and rhetoric/logic as a constant opening and mediation.[17] This style of metaxological thinking necessarily rejects duality in favor of interconnection and the movement back and forth between two coexistent creative factors.

Pickstock asserts an internal mediation of myth and logos through a subject oriented towards the Good or the praise of the divine. But it is wise to also note, alongside Rapp, that Plato's desire to master and enslave parts of the soul remains present despite the subsequent peace that is the product of this totalizing suppression.[18] Even if there is not an external suppression of another being implied here, Rapp would agree with Abram that the internalization of this is not enough to undo the hierarchy between the white and dark horses.[19] Mediation cannot fully remove the underlying repressive possibility in the metaphor of the soul, but it can suggest a less violent interpretation.

Unlike the dualist interpretation wherein the Good awaits us in the unchanging, perfected, and ideal realm of the Forms, Pickstock argues that the Good is articulated with the gifts it bestows, just as the sun can be said to be present in the light upon the rocks, the growth of vegetation, or the eye's reactionary squint in its rays. We remark that the sun is "in our eyes" when, of course, the sun is approximately 150 million kilometers away from our eyes. Pickstock notes that the Good is always "beyond being."[20] In Pickstock's reading, transcendence in *Phaedrus* is a plenitude that spills over into immanence and is revealed in physical particulars. Against the dualistic reading, Pickstock suggests that in *Phaedrus*, the love of dialogue is divine *and* it exists between two human beings: "The philosopher who, by the process of dialectic, has regained knowledge of the good, is wont to glimpse its transcendence in the mundane order, and is thus given to revere all physicality according to its participation in this spiritual sun."[21] In the process of dialectic, one engages in dialogue with other humans in an attempt to climb the ladder of love towards the Good, which is espoused in Diotima's speech in Plato's *Symposium.* Having glimpsed the Good through this process, its overabundance rains down upon one's experience of the material world and one witnesses the Good in all things. As such, transcendence necessitates immanence. Knowledge of the Good begins and ends in the world.

Rapp argues that body and soul are absolutely intertwined in *Phaedrus* and that this is found in the dialogue's rhetorical arguments, and it is performed in the men's interactions with one another and the natural world: "Socrates' speech and its word-images can be regarded as the bodily form of discourse through which Plato communicates ensouled logos." Rapp continues that the text demonstrates "entanglement between soul and body at the level of the surface content of the images and between ensouled logos and written text at the theoretical level."[22] Rapp marks the intermediaries emphasized in the text in both physical and theoretical ways. The soul and the body are intertwined, as are the mediums of logos and mythos. What truly overcomes any hard and fast mind–body dualism for Rapp and Pickstock is the performative layer of the ideal dialogue between the physical men Socrates and Phaedrus.

For Pickstock, the Platonic *theory of participation* in which all particulars participate in the universal Form frees *Phaedrus* from "otherworldliness" and the withdrawal from physicality. Participation allows for love of worldly, beautiful, particular things, not as a distraction from the Forms but as their embodiment. In Pickstock's reading, the Good is manifest in temporal reality and is not separate from it:

> [T]he eternal transcendence of the good which causes a kind of overflowing into physicality keeps us within the movement of time, since the philosophic access to the good arises not as something subjectively *a priori*, but rather in and through the forward-moving existence of what is given to be by the good. Such temporality does not compromise that knowledge, but rather constitutes its

> condition of possibility for us: the good arrives through time, and therefore time is not merely a ladder of access which can be kicked away, its job performed. For the philosopher makes no attempt to fetishize the present, to stem the flow of time, nor to forego the necessary narrative through time of his striving toward the good. Instead by embracing his narrative, he comes gradually to know himself, a knowledge coterminous with that of the realities towards which he strives. To this end, he inquires into every detail of the world, at once remote from and yet supremely incorporated within, his physical environment.[23]

The realm of the Forms in Pickstock's reading is a part of the temporal structure that constitutes the journey of life rather than a fixed spatial destination that completely detaches the individual from time and matter. When Plato refers to the Forms as "eternal," this points to the Forms as persistent and ever-presencing. In this way, the journey of the philosopher towards self-knowledge is never removed from her physical environment. One knows oneself and glimpses the Good of that self-knowledge in and through one's earthly experiences. For Pickstock, the two cannot be separate: "on account of the excessiveness of transcendence, the good is always overflowing into that subject which, via eros, strives to participate in it. A complete absence would suggest two discrete realms – the Forms and physicality – which for Plato would be an absurdity."[24] In Rapp's reading, knowing the self is a process that happens in circuitous cycles and this process both asserts and unmoors what we consider to be the self.[25] To know the self is not a stagnant position or a destination but a continually growing and changing hermeneutical path. It involves the moment of *metaxu*, which links the self to the divine; transcendence, which climbs the ladder towards the Good through epistemic engagement with the beautiful; and immanence, which recognizes the possibility of the participation of all things with the Good.

Plato and Nature

Unlike Plato's other dialogues, *Phaedrus* gives a rare spotlight to the natural world. Ferrari points out that in *Phaedrus* the two men cannot keep the background in its place. It is continually breaking into the foreground:

> what is particularly striking about this dialogue is that the background will not stay where it belongs. It becomes a prominent topic of discussion and a direct cause of the conversational action rather than, as one would expect, at most an indirect influence on its course. This in turn should prompt us as readers to scrutinise it more closely than we might otherwise have done.[26]

The relationship between Plato and nature has traditionally been viewed as one of denigration, as in the dogmatic reading of the text, or of possible indifference, as in the skeptical reading of the text. But contra these popular interpretations, Pickstock's non-dualist reading of *Phaedrus* blurs the lines between matter and ideal: "spontaneous adoration is inspired, via *eros*, by the memory of the beautiful itself which now infiltrates every aspect of life, great and small, in such a way that the 'ordinary' becomes the 'extraordinary.'"[27] Indeed, Pickstock even goes so far as to argue that while *Phaedrus* treats the physical particulars as a "mere backdrop, Socrates' holistic and harmonious understanding of interconnectedness disallows any such hierarchical demarcation of phenomena."[28] This assertion of Plato as absolutely bound to doxological interconnection between the ideal Good and the material beautiful is perhaps pushed beyond textual possibility in Pickstock's assertion that any Socratic disdain for nature is an exercise of Socratic irony alone, especially when we recall Socrates' assertion: "I am a lover of knowledge, and the men who dwell in the city are my teachers, and not the trees or the country."[29] This is a difficult line to read as purely ironic because Socrates truly does spend all of his other dialogues in the city in the company of men. However, as I have noted in the previous chapter, there are a number of paradoxes between rural and urban settings in the *Phaedrus* text. While I would not claim that Plato sides with nature in this dichotomy, I maintain that he works between the natural and urban realms in the *Phaedrus* dialogue, gesturing towards the possibilities in each distinct environment. Simply put, the division is never clear cut. It is varied, dynamic, and constantly shifting in a series of renegotiations.

Pickstock refutes any "anti-nature" themes in *Phaedrus* by claiming that Plato does not assert the interior/exterior distinction read into him by post-Cartesian philosophers. Because interior self-knowledge is gained through an exterior *eros* towards the material world with its beings and ideas, this ultimate Platonic goal of self-knowledge is a process of mediation and never simply the embodiment of one or the other side of a dialectic.[30] One cannot simply arrive in the realm of the Forms and stay there separated from matter. Socrates cannot disavow nature because it is a pivotal component of a balanced life. Pickstock notes that for Plato "genuine intellectual clarity is obtainable only when that which is to be 'known' is allowed to remain open and mysterious: an attitude synonymous with a kind of reverence."[31] This is precisely the necessary stance for the possibility of *metaxu* that I would like to assert in my work.

Ferrari, too, notes the practice of attention to the surrounding world in the *Phaedrus* and even suggests that Socrates' aim is to subtly instruct Phaedrus in the art of attention to the environment:

> when Socrates brings Phaedrus alive to his previously unacknowledged and unremarkable ability to orient himself appropriately in his surroundings, he does so without a professional context to frame and make a place for his extra-ordinary attention to the ordinary; and that accordingly, this attention seems out of place.[32]

Furthermore, Ferrari notes that Plato's epistemology is not propositional knowledge, but neither is it wholly unknowable or mysterious. This aligns with the suggestion of Scott and Welton that one comes to glimpse the Forms in and through the Socratic wisdom of ignorance. One can be directed towards the possibility of the Form through non-propositional knowledge. Plato's epistemology can be embodied through following the Socratic dialogue. For Ferrari, the possibility of hermeneutic *metaxu* is achieved by a continual shifting between background and foreground.[33] However, let us temper this with Rapp, who argues that Plato is a philosopher of impossibility whose goal is always to keep open mystery without solving or totalizing it: "emphasizing disruptive incongruity and the impossibility of resolution."[34] For Rapp, mystery always defies logical explanation. A common theme of these metaxalogical thinkers is allowing for mystery alongside logic.

The reading of Plato's *Phaedrus* from Pickstock is an interpretation of the subject as inherently doxological and as such, the praise of the divine Good is the foundation of ontological determination from the Good, epistemic return to the Good, and ethical actions in the material Good of the world. This practice of hymn and praise is an aesthetic act, which in itself is contrary to the infamous suggestion of Plato that artists should be banned from his Republic.[35] Pickstock notes that "Socrates' indictment of the poets is therefore not a condemnation of poetry as such, but rather of the separation of language from doxology, of art from liturgy, resulting in a sophistic 'virtual reality,' or realm of mere fiction which is manipulable, ironic, and uninhabited."[36] The condemnation is of poetry that has forgotten its place as a connection to the Good. Pickstock sees the Platonic subject as striving to attentively and openly praise the Good through an erotic love and quest for self-knowledge. This is founded on action based in the material world and co-created by body and mind, which cannot be understood to be distinctly separate but rather must be co-created and co-creative.

Weil's *Metaxu*

Like Pickstock's doxological reading of Plato, Simone Weil's work offers a distinctly mystical interpretation of Plato's dialogues. In a letter, Weil notes that "ecstasy" is the proper state of mind for reading *Phaedrus*.[37] Weil's interpretation of Plato is described by Emmanuel Gabellieri as a profound counterargument to the history of the western tradition as well as contemporary philosophy: "Simone Weil's thought is of major significance not only for contemporary philosophy but also for the history of philosophy because she offers a counter argument, and perhaps the most coherent and profound alternative, to the program of overthrowing Platonism."[38] While so many post-modern thinkers wish to overthrow the Plato of dualism and exclusion, Gabellieri directs us towards Weil if we seek another path forward. He goes on to describe Weil's Platonism as founded on her reading of *metaxu*: Weil's "is a philosophy of the constituent bond between reason and love, a philosophy of being as *metaxu*, link, mediation, and finally, a philosophy of the fullness of being as supernatural love."[39] Similarly, in his work on Weil's use of Pythagorean harmony, Eric Springsted draws on the "metaphysics of mediation"[40] in order to bring together her "rational" concrete solutions to the daily problems of laborers and her "spiritual solution to the problem of labor."[41]

Weil reads Christianity back into ancient Greek texts in a creative interpretation that suggests the truth found in Christianity is not unique to Christians but can be found throughout human history.[42] As Marie Meaney suggests, Weil acts as an apologist for Christianity to atheistic academics and an apologist for ancient Greek philosophy and tragedy to Christian thinkers and theologians. Weil chooses Truth above God, but she also believes that following truth leads one to God (*AL*, 58–76). According to Meaney, this Christological interpretation of the ancient Greeks is an example of *metaxu*:

> Weil wants to incite her readers to walk over the bridges contained in the Greek classics towards that other reality "outside the world, that is to say, outside space and time" (SE 219); which is a person, namely God, "who holds us in his arms from the beginning" (SL 137).
>
> (*AL*, 76)

Thus, an understanding of *metaxu* in Weil's writings on Plato must take into account a specifically Christian reinterpretation of Plato's concept. But rather than seeing Weil as inserting Christianity back into pre-Christian ancient Greek texts, which would seem nonsensical, Weil is claiming to find a shared allegiance to truths that might be mutually enlightening. It is also notable that there are many traditions that Weil draws from and while Christianity is dominant, it has perhaps been given more emphasis than Weil would have wanted.

Learning to Die

If, as is suggested in *Phaedrus*, Plato's ultimate goal is self-knowledge through philosophy, this should be considered alongside the Platonic suggestion in *Phaedo* that philosophy is a training for death, or according to Michel de Montaigne, to philosophize is to learn to die.[43] Weil claims that "Plato does not say, but he implies, that to become wise, which exacts the knowledge of self, one must become, already in this life, naked and dead. The examination of conscience exacts this breaking of all the attachments which make up our reasons for living" (*IC*, 82). Self-knowledge as mediation between logic and myth is also a mediation between life and death in Weil's reading of Plato. For Weil, the act of philosophy means learning to die and it involves the decreation of the I and the other as objectively separate entities. Decreating the "I" holds one out over the void of existence and denies one the reliance upon the tunnel-visioned future-orientation of individuated subjectivity. One is denied the future as a guaranteed certainty and instead must accept the inevitability of death and release the attachment to the future with love and acceptance:

> Only the true renunciation of the power to think of everything in the first person, the renunciation which is not a simple transference, grants a man the knowledge that other men are his fellows. This renunciation is the love of God, whether or not the name of God be present in the mind.
>
> (*IC*, 175)

To accept the inevitability of death and to give up the first-person conception of the world makes ethics possible and sets one on the path towards God. Ethics and *amor fati* here on earth direct our desire towards the Good beyond the earth. Here we see a distinction between Weil and Pickstock: whereas Pickstock maintains that the Platonic Forms operate *in* time, Weil maintains the status of the Forms as both in time and *also* transcending time in eternity:

> *The Phaedrus*. The beauty which is on the other side of the sky, here it is, visible to us. Plato does not say how it has arrived here below. Obviously, it must have come down. There must therefore be a descending movement unconnected with gravity, which is love (what else could it be?) … In everything which rouses in us a pure and genuine feeling for beauty, God is really and truly present. There is, as it were, a sort of incarnation of God in the world (*Timaeus*) of which beauty constitutes the sign.
>
> (*NB*, 440)

It is through the love that retracts divine being and allows for creation that divine being is present in its very absence. To describe this presence-in-absence, Weil uses the analogy of feeling the presence of a deceased loved one through the pang of her absence:

> To lose somebody: we suffer at the thought that the dead one, the absent one should have become something imaginary, something false. But the longing we have for him is not imaginary. We must go down into ourselves, where the desire which is not imaginary resides … The remedy is to use the loss as an intermediary for attaining reality. The presence of the dead one is imaginary, but his absence is very real; it is henceforth his manner of appearing.
>
> (*NB*, 28)

This act of experiencing one's grief and loss as an intermediary does not suggest a removal of the self to the realm of Forms or the eternity of the soul beyond the material world, but rather, it helps one to comprehend *reality*. Shifting this idea from an absent beloved human to the absence of God, there is an opening for the descent of love: "God's absence is the most marvellous testimony of perfect love, and that is why pure necessity, the necessity which is manifestly so different from good, is so beautiful" (*NB*, 403). God makes the divine known to us through the beauty of the world. In fact, the only way that God can make the divine known to humans is through material mediation. In this act, we recognize that the material and the spiritual are interwoven. Such a reading places the Platonic Good as a realization in the material world and not a rejection of the material world. But it also maintains the mystery *beyond* the material world. The Good is simultaneously both *outside* time and space and *in* time and space.

Gabellieri argues that the interconnection of Platonic transcendence and the descending movement of the Good into immanence is considered as all-important in Weil's reading of Plato.[44] Gabellieri notes that the dimension of revelation surpasses pure reason and acts as the transcendental lynchpin for Weil's grace or the Platonic experience of transcendence. In Weil's reading, the Platonic dimension of love favors the non-intellectual (ethical and mystical) path over the intellectual (rational and dialogical) path.[45] Springsted observes: "The doctrine of love which Weil finds in the *Symposium* provides this bridge, for love which descends from God and is engendered in human hearts is discussed in that dialogue as being that which mediates between Need and Plenty."[46] To forget this element of love is to miss out on the pivotal descent of God to the earth and the dynamic connections between the human and the Good.

Gabellieri claims that Weil sees *metaxu* as the recognition of *relation*, of the movement "in-between" the physical I and the other infused with Truth or the Good. To grasp an idea is to reify it and hold it in an unmoving place (the separate realm of the Forms), whereas to recognize a relation is to recognize that one is ethically bound to another being.[47] The relational nature of *metaxu* in Weil embraces the reality of death. Enigmatically, her notebooks state: "To think on relationships is to accept death … Core of Platonic thought" (*NB*, 145). Increased understanding of our relationality involves an increased vulnerability both in our own mortality but also and more importantly in the mortality of those around us. It is a recognition of ethical duty.

For Weil, decreating the self through the cultivation of attention and the subsequent recognition of *metaxu* is the process of the philosopher freeing herself from the Platonic cave. But this freedom comes about in and through the cave and of course, the philosopher must return to the cave. In Weil's reading of Plato, to exist in the cave as a place of shadows and illusions is not to be condemned to a meaningless attachment to the things of the material world:

> The unreality of things, which Plato so powerfully depicts in the metaphor of the cave, has no connection with the things as such; the things in themselves have the fullness of reality in that they exist. It is a question of things as the object for love. In this reference they are like shadows cast by puppets.
>
> (*IC*, 134)

Rather than reading the shadowy cave as a denigration of the natural world and matter, Weil directs us to consider the allegory as a way of understanding the possibility of love. Our love towards the world is weak until we become attentive, allow the world to be, and see the reality of the sunlight. This is not a separate realm of existence but a new way of loving. In such a reading, the "great beast" of society creates these shadows (*NB*, 618). Gabellieri notes that a consequence of Weil's Christian Platonism is that she does *not* conclude with an escape from the world, but with an orientation towards incarnation. One must open up to the reality of this earthly life because without attention and decreation, we encounter only shadows and dreams.[48]

Bridges and Walls

Weil defines *metaxu* as "bridges" (*NB*, 370), but they can also be walls (*NB*, 497). They are passageways or roadblocks but not destinations in and of themselves.

It is only through supernatural love that humans can experience *metaxu* as these momentary process-oriented paths and not attempt to reify them as absolutes. Weil argues that "every separation represents a bond" (*NB*, 497), which asserts an ambiguity to all things: there is always the possibility that something can act as *metaxu* but also that it can lead one into illusion, an *ersatz metaxu*. All created things and temporal moments or events are potentially these *metaxu*, which can both open links and enforce separations.[49] For Weil, these things link to one another and also to God. God, or the Good, speaking in more Platonic language, is the only possible end. Everything else is simply means. It is for this reason that *metaxu* are the only truly sacred experiences we can access, albeit fleetingly. There are many vulgar pursuits in life that are shallow, misplaced, *ersatz* ends and as such are not divine. On the other hand, God, as the only true end-in-itself, is beyond our reach and even our conceptualization. Weil claims that *metaxu* alone exists in our range of possibility as that which form "the region of good and evil." And Weil goes on to describe:

> it is useless to try to raise oneself above the μεταξύ, one remains in the sphere of good and evil through the relations established with those of the other people who are there too. It is impossible, therefore, for the problem of good and evil to disappear in the movement of ascent.
>
> (*NB*, 48–49)

It is this concept, taken directly from Plato's *Symposium* by Weil and put to work as foundational in her own philosophy, that suggests a non-dualist reading of Plato. She even goes so far as to posit that the concept of specialization in Plato's *Republic* is "specialization of the faculties in man and not of the specialization of men themselves; the same applies in the case of hierarchical order" (*NB*, 447). According to Weil, when Plato speaks of specialization and of hierarchy, he is not referring to an external differentiation between humans but rather of an internal process of *metaxu* in which specialization and hierarchy of one's faculties are cultivated.

One way that Weil describes *metaxu* is as a stick to a blind man, a metaphor utilized by Descartes. Andrew Davison explains: "This learning to sense God in things – learning to feel with the world as a blind man feels with his stick – takes practice. These sticks, what she calls bridges or *metaxu*, are many and various."[50] The *metaxu* become a prosthetic limb, enhancing our spiritual senses of the Good and allowing us to experience that which we could not experience with our individual senses alone.

Mystical Hermeneutics

According to Weil, Plato is a "mystic, heir to a tradition of mysticism wherein all Greece was bathed" (*IC*, 74). As such, the wisdom of Plato is an "orientation of the soul toward grace" (*IC*, 85). Weil considers the foundation of Greek culture and history to be steeped in remorse for their actions against Troy during the Trojan War and as such, they are a people overwhelmed with the desire to comprehend and contemplate misery and force: "All Greek civilization is a search for bridges to relate human misery and divine perfection. Their art, which is incomparable, their poetry, their philosophy, the sciences which they invented (geometry, astronomy, mechanics, physics, biology) are nothing but bridges" (*IC*, 75). Weil is absolutely clear that *metaxu* is the foundation of the Greek world and that this *metaxu* is born of a sober attention to the very real human suffering that the Greeks perpetrated during the Trojan War. This openness and attentiveness to humanity allows one to distinguish between what is an *ersatz* ethical demand of the "great beast" that is society and a true ethical demand of God or the Good.

Metaxu are the grace that we receive when we are open and practicing philosophy as the quest for self-knowledge through an attention to others, to the world, and to God. Weil notes that readers can receive this grace from the images created in texts, and she suggests *Republic*, *Phaedrus*, and *Symposium* as particularly apt examples: "The fundamental idea of these images is that love is the disposition of the soul to which grace is given, which alone is able to receive grace, love and none other than love. Love of God is the root and foundation of Platonic philosophy" (*IC*, 88). In conveying a mystical truth, a text itself can act as *metaxu* if the reader adopts the proper attentive ecstatic attitude.

Metaxu emerge through the cultivation of knowledge, love, and particularly in attention to the beauty of the world: "beauty is the only value that is universally recognized" (*IC*, 103). For Weil, the beauty of the world speaks to us and offers us a link to the divine. But this beauty is not God Itself. Rather, it is as the presence of the deceased beloved brought forward through their staggering absence. It is in this moment of presence-in-absence that we can comprehend the nature of the love between ourselves and our deceased beloved; and so too, it is in the beauty of the world that we can see the absent-presence of God, feel that absence to our very core, and be both overwhelmed and inspired by the nature of our relationship to the Good, to the divine, as saturated in love. Reflecting on *Timaeus*, Weil notes the cyclical nature of becoming both in the Platonic cosmology and in the Platonic quest for self-knowledge:

> Plato is thinking not only of the cycles of the day, the month and the year, but also of the ideas which unite them in his system of symbols, that is the Same and the Other; in other words, identity and diversity, unity and multiplicity, absolute and relative … in these contrary pairs, which make but one, the second term is not symmetrical with the first but subordinate, even while being opposed to it. (*IC*, 98)

This speaks again to the constant process of *metaxu*, of a mediation that never rests but continually becomes and grows.

Turning to Socrates' speech in *Symposium*, Weil notes that "there is sufficiently clear indication that the doctrine contained in this work does not issue from philosophical reflection but from religious tradition" (*IC*, 124). Socrates leaves the city and enters into the sacred grove with Diotima in order to learn of love. And of course, this is where Diotima defines love as an "in-between," as something above man and below God: "It [Love] designates the mediators, the intermediaries between man and God" (*IC*, 125). Love is the desire of the "in-between" when we experience the bridge or the passage to the Good. Weil reads this as a harmony between the contraries of God and Want.

Dualistic readings of Plato are, according to Weil, absolutely mistaken, because Plato is not interested in a division of Forms from earth but rather "it is a question of a spiritual marriage with the beautiful, by the grace of which the soul truly begets virtues" (*IC*, 145). The desire to reach towards the ideal Good is not a movement away from the world to an abstract realm beyond the world but rather a spiritual journey that finds inspiration in the beauty of the world. For Weil, this comes out of the Pythagorean sense of harmony, which she sees Plato as steeped in, embracing

> the Pythagorean idea of harmony as the union of contraries, and the combination of that which limits and of that which is limitless ... To him who lovingly contemplates the order of the world, there shall come a certain day and moment when suddenly he shall contemplate another thing, a miraculous sort of beauty. (*IC*, 148)

Gabellieri notes that in Weil's reading of *metaxu,* Plato

> allows us to conceive of a mode of being beyond the alternative between immanence and transcendence and also beyond the dilemma of presence and absence in which the phenomenological critic of metaphysics risks becoming entrapped ... The path undertaken in *Philebus* that Weil makes her own: to develop a philosophy of revelation that avoids "reaching the One too quickly" and forgetting that, since revelation does not suppress mystery, "what is hidden is [always] more real than what is manifest" (C II 273–74).[51]

That which is hidden, the love of God and our desire for the Good, is the reality that is both immanent in the material world and also transcends any attempt to capture it or pin it down. When we can see the world's beauty as infused with a love and an ethical duty that are neither reducible to material things nor separable from them, then we are seeing the world in its reality. Weil's reading is a non-dualist interpretation of Plato that allows for a becoming and opening, which never completely rests in a stagnant or reified realm but continues to grow in self-knowledge and love through ethical actions in the world.

Notes

1 Gary Alan Scott and William A. Welton, *Erotic Wisdom: Philosophy and Intermediacy in Plato's* Symposium (Albany: State University of New York Press, 2008), 19.
2 Ibid.
3 Ibid.
4 Ibid.

5 Ibid., 22.
6 Ibid., 25.
7 Ibid., 23.
8 Ibid.
9 William Desmond and Kevin Kennedy, "The Theater of the *Metaxu*: Staging the Between," *Topoi* 30.2 (2011), 113.
10 Scott and Welton, 23.
11 In his speech about love, Aristophanes tells a tale of the original humans being round creatures composed of what we would now consider two separate humans. These creatures became too powerful, so the Gods cut them down the middle. In this myth, love means to reconnect with the person who is literally one's other half. See Plato, *Symposium*, ed. M.C. Howatson and Frisbee C.C. Sheffield, trans. Howatson (Cambridge: Cambridge Press, 2008), 189b–193e, p. 22–27.
12 Scott and Welton, 23.
13 Catherine Keller, *Face of the Deep: A Theology of Becoming* (New York: Routledge, 2003), XVIII.
14 Catherine Pickstock, *After Writing: On the Liturgical Consummation of Philosophy* (Oxford: Blackwell, 1998), 4.
15 Ibid., 5.
16 Giovanni R.F. Ferrari, *Listening to the Cicadas: A Study of Plato's* Phaedrus (Cambridge: Cambridge University Press; 1987), 33–34.
17 Ibid.
18 Jennifer R. Rapp, *Ordinary Oblivion and the Self Unmoored: Reading Plato's* Phaedrus *and Writing the Soul* (New York: Fordham University Press, 2014), 7.
19 David Abram, *The Spell of the Sensuous: Perception and Language in a More-Than-Human World* (New York: Vintage Books, 2017), 122.
20 Pickstock, 12.
21 Ibid., 12.
22 Rapp, 12
23 Pickstock, 13.
24 Ibid., 22.
25 Rapp, 10.
26 Ferrari, 3–4.
27 Pickstock, 16.
28 Ibid.
29 Plato, *Phaedrus*, trans. Benjamin Jowett (Bulgaria: Demetra Publishing, 1870), 230d, p. 10.
30 Pickstock, 29.
31 Ibid., 16.
32 Ferrari, 16.
33 Ibid., 23.
34 Rapp, 92.
35 Plato, *Republic*, trans. G.M.A. Grube (Indianapolis: Hackett, 1874), Book III, 398a, p. 68 and Book X, 595a–608c, p. 241–252.
36 Pickstock, 42.
37 Simone Weil, *Seventy Letters*, trans. Richard Rees (Oregon: Wipf and Stock, 1969), 86.
38 Emmanuel Gabellieri, "Reconstructing Platonism: The Trinitarian Metaxology of Simone Weil," in *The Christian Platonism of Simone Weil,* ed. E. Jane Doering and Eric O. Springsted (Notre Dame: Notre Dame Press, 2004), 133.

39 Ibid. Gabellieri extracts this Weilian reading of Plato from her texts focusing on Christian Platonism, which Weil composed during her time in Marseille: "God in Plato" and "Divine Love in Creation."

40 Springsted draws on Weil's essay "The Pythagorean Doctrine" for this reading.

41 Eric O. Springsted, "Chapter IV: Mediation in Daily Life," in *Christus Mediator: Platonic Mediation in the Thought of Simone Weil* (Chico: Scholars Press, 1983), 195–196.

42 For example, in the essays posthumously compiled in *Intimations of Christianity Among the Ancient Greeks*, Weil examines Greek myths, tragedies, the Presocratics (in particular the Pythagoreans), Homer, and Plato.

43 Plato, *Phaedrus*, 229e, p. 9; Plato, *Phaedo*, trans. David Gallop (Oxford: Clarendon Press, 1975), 81a, p. 29; Michel de Montaigne, "Chapter XIX: That to Study Philosophy Is to Learn to Die," *Essays of Michel de Montaigne,* trans. Charles Cotton, ed. William Carew Hazlitt (Project Gutenberg, online EBook #3600, 2006).

44 Gabellieri, 143.

45 Ibid., 144.

46 Springsted, *Christus Mediator*, 208.

47 Gabellieri, 146.

48 Gabellieri, 136.

49 Many things can be read as *metaxu* in Weil. Consider, for Example, Meltzer's suggestion that hands act as *metaxu*. Françoise Meltzer, "The Hands of Simone Weil," *Critical Inquiry* 27.4 (2001), 624. And Springsted points to a limitation in our knowing as a *metaxu*: Eric O. Springsted, "Of Tennis, Persons and Politics," *Philosophical Investigations* 16.3 (1993), 206.

50 Andrew Davison, "The Mediating Possibilities of Absence in the Thought of Simone Weil," *Theology* CXII. 865 (January/February 2009), 11.

51 Gabellieri, 151.

Part III

Decreation for the Anthropocene

8 Weil and Anthropocene Ethics

In the first two parts of this book, I have established an ethical map for reading Simone Weil and considered her non-dualist interpretation of Plato against traditional dualist readings. In this final part of the book, I will argue that Weil's notion of *decreation*, as it has been explained through non-dualist Platonism, is a potent method of thinking and acting through what some have deemed the current geological epoch of the Anthropocene. Following a critical examination of the multiple meanings of the term Anthropocene, this chapter will situate Weil's philosophy within these debates and make explicit connections to deep ecology, a term coined by eco-philosopher Arne Naess, to illustrate how Weil's philosophy is ecologically potent. Through this engagement with the concept of the Anthropocene and Naess' deep ecology, I will establish a Weilian-inspired ecological ethics as a relational interconnection and responsibility to the impersonal good in all other unique beings while maintaining a priority for the responsibility for fellow human beings. For Weil, this is a humanist ethical project that necessitates social justice, and for my own work, this ethical interconnection, responsibility, and justice extend to all beings. In the ninth chapter, I will unpack potential issues in expanding Weil's thought into an ecological ethics of predominately non-hierarchical interrelation, namely, issues around humanism, God, and purity. All of these notions are undeniably present in Weil and suggest a hierarchical anthropocentrism, which I will not completely dispel, but will temper. Just as a non-dualist reading of Plato yields ecological insights, so too a critical non-dualist reading of Weil yields ecological insights with a strong emphasis on social justice. Weil's understanding of these potentially problematic concepts is founded upon her non-dualist reading of Plato, and as such, maintains the possibility of an ecological ethics. In the final chapter, I will turn to specific examples of how one may ethically act ecologically in light of Weil's ethics. This Weilian-inspired ecological ethics for the Anthropocene involves a recognition of the radical depth of responsibility, the importance of social justice, and the sense of loss that is at stake when we are fundamentally co-constituted through others, both human and non-human.

DOI: 10.4324/9781003449621-12

Towards a Decreation for the Anthropocene

Viewed as the period during which human activity has been the dominant influence on climate and the environment, the name "Anthropocene" was suggested in 2000, at a conference in Cuernavaca, Mexico, when Paul Crutzen, a civil engineer, meteorologist, and atmospheric chemist, suggested that the title aptly characterized an epoch of human dominance in biological, chemical, and geological processes on Earth.[1] Anthropogenic changes, like those caused by climate change, will radically change the future in an exceptional way akin to previous epochal changes marked by geological periodization. The Anthropocene is often considered to have begun in the mid-twentieth century with the exponential rise of resource exploitation after the Second World War termed the Great Acceleration,[2] although some, including Crutzen, date its origin further back to the start of the Industrial Revolution. As we shall see, this claim is ostensibly one based wholly in the natural sciences, but as many have pointed out, the epoch, however defined, has been specifically associated with and fueled by capital-driven growth and an associated obsession with accruing more and more material wealth. As I work on this book, across the province in Milton Ontario, Canada, Crawford Lake's well-preserved layers of sediment are being studied by pulling and analyzing ice cores. Due to its unique depth and climate, this lake contains evidence of the Great Acceleration, and as such, in the coming years, scientists may establish an exact timeline of the Anthropocene. But as I write, the term remains open.

The period associated with the increasing global dominance of capitalism and its environmental consequences is also marked by the ascendance of culturally and historically specific ideologies of human exceptionalism and individualism.[3] These ideologies presume an understanding of humanity as radically separated from and increasingly dominating nature, and of individuals as autonomous agents separated from nature and their fellow human beings. These ideological presumptions are also, ironically, echoed in the notion of the Anthropocene itself, which is why Clare Colebrook, for example, suggests that those critical of Anthropocene discourse must not just "accept the Anthropocene as an epoch, a line or strata whose significance is not in dispute … rather we might ask for whom this strata becomes definitive of *the human*."[4] The very notion of the Anthropocene emerges out of a modern western worldview of a humanity that is regarded as separate from, and sets itself above, the non-human world. This means that we have to carefully interrogate the complex history of terms like "humanity" and "nature" and their role in modern western society, philosophy, and environmentalism.

The ideological "individuation" of the human subject, rending it from the larger social, political, and physical ecologies of which the human is an integral part, may also be a contributing factor to the alienation and destruction that has befallen our era. If this is the case, then in order to move forward ecologically, the individuated human-as-capitalist-consumer must be reimagined. This rethinking of the individual is the first way in which decreation is relevant to the Anthropocene: in my expanded account of Weil's concept, decreation involves the individual act of decreating the "I" in favor of seeing oneself as a part of a larger ecology while mindfully allowing for other aspects of that ecosystem to emerge. This first form of decreation is a personal spiritual journey of self-knowledge, interconnection, and responding to the difference of more-than-human others. This decreative shift is not only ethical, but also ontological and epistemic: it is an alteration in our *thinking* to rediscover our *being* as both singular and plural. This will be the main focus of this final part of the book although we will also have recourse to the argument presented in Chapter 3 regarding the importance of understanding the tension and balance Weil posits between force and necessity, since this speaks directly to core aspects of the Anthropocene.

The second form of decreation that is required in the Anthropocene is therefore the reimagining of the very epoch itself. The name "Anthropocene" reasserts, even if it partially critiques, the consequences of human dominance over and against nature; it places human beings in the realm of culture and objectivity, commanding the natural world, molding and altering it to suit their own needs and selfish desires. In this sense, it reinforces the very narrative that this book seeks to overturn. Like Colebrook, many critics including Donna Haraway, Jason Moore, and Anna Tsing have problematized the term Anthropocene for this very reason. Moore offers the alternative term "Capitalocene," which recognizes that it is actually an incredibly small number of humans who benefit disproportionately from the capitalist economic system destroying the natural world, and it is not human beings as a whole that are the problem.[5]

A further issue with the term Anthropocene is that it maintains the human as the all-important center of the world. To avoid this anthropocentric vision of the Anthropocene and the despair associated with living in a Capitalocene, Haraway suggests the term "Chthulucene," which is derived from *chthon*, meaning "earth" in Greek, and is associated with things that dwell in or under the earth. The Chthulucene, for Haraway, refers to processes of re-worlding. She suggests it is a process of composting, of allowing that which is no longer flourishing to serve as the fertile grounds for radically new and different growth to emerge. We cannot suppose that which emerges from our present situation will reprise or sustain our current social ecologies and philosophies and instead we must remain open and attentive to the possibility of the radically new. Haraway suggests that while the definition of "human" may change in the process, it is not a matter of being "posthuman" but rather of allowing for openness, attention, decay, and new growth.[6]

Moving beyond this epoch must entail a cessation of propping up dead structures or supporting destructive conditions for human and non-human life, such as unbridled capitalism. Tsing notes that in tying the human condition to capitalism, we have problematically entangled the human narrative with that of progress, alienation, and seeing all other beings as resources. For Tsing, the concept of the Anthropocene both reinforces and attempts to overcome this capitalistic and colonial conceit and so, although not wishing to discard the term Anthropocene altogether, she also suggests the title of the "plantationocene," which emphasizes the colonial commodification of racialized humans and non-human beings through the establishment of the Southern American plantation. Like capitalism, slavery and the American plantation undermined the conditions of life on earth for many to the benefit of very few.[7] However, Tsing finds that violence, extinction, and endings can also bring about new growth and possibilities, one of her key examples being the growth of matsutake mushrooms after the atomic bomb dropped in Hiroshima. From a destructive violent act, life returned and grew anew, and from a recognition of our past mistakes and misunderstandings about our place in the world, potentially fruitful ways to look forward might also emerge.

Similarly, David Farrier maintains the importance of holding open all of these potential terms and does not want to place one term hierarchically above any other but wishes to recognize "each as a useful prism for viewing contemporary crises."[8] However, perhaps in order to be helpful in recognizing some of the issues we now face, the term Anthropocene itself must be decreated. It must be composted in order for new possibilities to grow from its rotting concepts. Catherine Keller notes that the epoch of the Anthropocene has not shown humans the limit of our impact on the earth, but rather the limit of our control over the earth.[9] In Weilian terms, the overly *forceful* rearrangement of ecologies and societies to suit the powerful has run up against earthly *necessities*. The destruction of the planet is not so much a matter of human domination of the earth as it is the inability of humans to fully control the earth. As many other cultures have long recognized, humans are limited in our power to control and dominate nature. We cannot wield force forever. The current ecological issues or the Weilian necessity that we now face reveal the limit of human domination and the impossibility of an exceptionalist understanding of our relation to the "natural" (in the sense of always already more-than-just-human) world. The subtitle of this book, "Decreation for the Anthropocene," is intended to recognize the limits of the term problematized by Keller, Farrier, Tsing, Haraway, Moore, and many others, and thus suggests that the Anthropocene itself needs to be decreated, to abnegate its human-centered, subject-oriented ethos. These concepts of human exceptionalism and alienation are, at least in part, derived from the division of "nature" from "culture" and as such, I will turn to the concept of "nature" as it has been set up in opposition to "culture" and then explore how Weilian ethics can help us think through these concepts by looking to the overlap between deep ecology and decreation.

Anthropocene Natures

It is important to briefly address the question of "nature," because one of the key aspects of the Anthropocene, raised by both its proponents and many of its critics, is that it seems to spell the end for any concept of nature as something entirely separate from human activities. Yet it is this concept of nature, for example, in the form of nature as romanticized people-free "wilderness," or nature as essentially opposed to humanity, culture, or artifice, that is often, rightly or wrongly, associated with various forms of environmentalism and environmental ethics.[10] The Anthropocene, Henrik Ernstson and Erik Swyngedouw claim, "affirms that humans and nature are co-produced and that the particular historical epoch that goes under the name of capitalism forged this mutual determination." The term

Anthropocene is "just another name for insisting on Nature's death," where Nature is "framed as the imaginary externally conditioning sphere for human existence."[11]

The impossibility of finding an immaculate Nature entirely free of human influence is, of course, something many environmentalists had recognized well before the term Anthropocene was established.[12] We also need to be very clear about exactly which understandings of the term nature are undermined by discourses claiming the "death of nature." In some cases, these discourses around the death of nature are, ironically, associated with forms of human ethical exceptionalism, which employ a distinction that regards nature as no more than a human resource. As Kristin Asdal notes, the western humanities tradition has long "restricted its focus to assumed autonomous human actors in isolation from nature and the material world."[13] Nature is set up as an object over and against the active human subject. This tradition has left "nature" as an objective fact for the natural sciences and has not, until recent years, turned explicitly *social* theory or the *humanities* upon the natural world. Applying the humanities and social sciences to nature complicates the possibility of nature's agency, relationships, and exchanges with the "culture" it has traditionally (in western approaches) opposed, but it does not necessarily mean that nature is dead.

Following Weil's critique of science, we cannot seek a final answer as to what nature is from a positivist science alone. Indeed, Asdal notes that when we take nature into account in the face of science and scientific theory, we can find ourselves asking "which science" it is that we should turn to in studying nature. "Feminist philosophy of science, for example, has posed fundamental questions about what we at any given time perceive as science and nature."[14] Feminist perspectives note that the divisions we make in these categories of "science" or "nature" can be arbitrary and that they are distinctly different depending on one's historical social context. Tsing notes the differences of presuppositions in Japanese and American science when it comes to the study of matsutake mushrooms.[15] Farrier notes that the urge to see distinctions such as "nature" versus "culture" is "rooted in an urge to simplify – to impose an order and regulate what belongs and does not belong."[16] If we ignore the details and interconnections of beings, we can exist in the black-and-white world of nature and culture, but the leaky, interconnected earth and all of its beings are ultimately where ethics must be enacted. There is nothing simple about "nature" in itself, and when we impose simplicity upon nature, Farrier argues that we create the sacrifice zones of environmental racism, extinction, and brutal environmental degradation. This distinction can sow radically unjust forms of environmental ethics and cover over the unjust acts. Before even considering the non-human beings in this epoch, Farrier bluntly observes, "in the Anthropocene, some human lives are deemed expendable."[17] If nature is simply a backdrop to human affairs, we are free to exercise our will to do to it what we please. And yet, despite the nefarious implications of this distinction, there is a necessity to refer to "nature" from time to time, because a broader understanding of nature, expanded to include human activities and concerns within the more-than-human world, is, as I have observed in Weil, the realm of both force and necessity: energy, power, and the possibility of ethics and the Good are all part of the realm of "nature" in this broader sense. It is, we might say, both the material matrix of our inter-relationality and a conditioning sphere for human existence.[18] These inter-relations are not confined to human use or natural resources but matters of concerned involvement in myriad ways. The way I will use the term nature moving forward recognizes something of this complex history, in which Weil herself, and any attempt to apply Weil's thought to ecology, is inevitably entangled.

"Nature" can hold as a concept but only as a leaky concept that admits to its own problematic encounters within and beyond its domain. The *Oxford English Dictionary* defines "nature" as "The phenomena of the physical world collectively; *esp.* plants, animals, and other features and products of the earth itself, as opposed to humans and human creations."[19] But this definition of nature is the eighth given, and preceding it are definitions including excrement, semen, menstrual discharge, female genitals, sexual impulses, urges, vital functions of the human body, the need to defecate or urinate, the "inherent dominating power or impulse in a person by which character or action is determined, directed, or controlled," and "Chiefly Christian Church. This impulse as contrasted with the perceived influence of God on man."[20] Thus we can immediately see that etymologically nature is never *simply* nature. It is a whole host of ideas, theories, and structures of power with an emphasis on baseness, materiality, excrement, femininity, and an absolute opposition to the Good. To think of nature as somehow apolitical or distinct from culture is nonsensical. Built into the western history of the word, nature is its implied opposition to culture.

The distinction between "nature" on one hand and "culture" on the other hand has been problematized by a number of ecological scholars. As Bill McKibben notes, "An idea, a relationship, can go extinct, just like an animal or a plant. The idea in this case is 'nature,' the separate and wild province, the world apart from man to which he adapted, under whose rules he was born and died."[21] The Anthropocene is, in one sense, a recognition that the impact of human action in and on nature is simply too intense for humanity to be able to distinguish itself from the natural world. Michael Smith describes moments when something is obviously both natural and cultural: "a rice paddy, an agricultural field, a city park, a weather system affected by global warming."[22] Or consider a whale fall in the abyssal depth of the ocean, deceased from an overabundance of ingested plastics. As Farrier notes, "Plastic typically appears to us as disposable, decoupled from its contexts of extraction and production, yet single-use plastic may remain a lively influence on marine environments for tens of thousands of years."[23] The human-made plastics that killed the whale are the formation of this new abyssal ecosystem. This is an event that is both natural and cultural: the material and the good, or rather, the good (and bad) in the material. The lived experience of nature and culture cannot always (or perhaps ever) be cleanly divided, and as such ecology is not simply about "nature" but also about social theory and the realm of culture. Smith explains: "Recognizing the reality of social relations does not mean that we are driven to deny every society's dependence upon a more-than-social world that predates it and will also antedate it (albeit in forms that have often been irremediably altered through these encounters)."[24] Social relations and realities are in addition to environmental necessities. For Karen Barad, nature and culture must be thought "without defining the one against the other and holding either as the fixed referent for understanding the other."[25] The idea of nature and that of culture can be parsed as distinct intellectual realms, but the lived experience of each is somewhat messier.

It is admittedly difficult to avoid falling back into a nature/culture dichotomy even when trying to critique its consequences. Robert Harrison argues that people are conflicted in their own desire for and repulsion from nature. For Harrison, we view nature, on the one hand, as a sanctuary from the demands of autonomy and the obligations of the polis. Nature is a respite from the responsibility of cultural life, offering a simplicity and deep connection. On the other hand, Harrison argues that nature carries with it a sense of the uncontrollable and abject aspects of human life.[26] Nature is the harbinger of necessity (in the Weilian sense) and rather than a respite, it can be viewed as a prison that robs us of the freedom of cultural civilization, demanding a pure material survival against the elements. This paradoxical relationship towards nature manifests with particular voracity in the neoliberal capitalist state or the human action that desperately attempts to recreate a serene, naturalized terrain. There is a schizophrenia cultivated in this divided lust and repulsion towards "nature." This is fueled in no small part by capitalism and property rights going back to John Locke, who advised men to mix their labor with the land in order to make that land rightfully their own and instill value into what had been inherently a blank slate without value of its own. Harrison contends that humanity's frenzied consumerist desire both wants the child-like eternal state of the garden of Eden and simultaneously pushes back against the bestial removal of our worldly political and relational actions that such a garden would bring about: "One of the perverse consequences of this paradox is that the path back to Eden is littered with ruins, corpses, and destruction. Our attempts to re-create Eden amount to an assault on creation. That is the danger of the era."[27] In our desire to control and possess nature as a place of solace, we inadvertently destroy both nature and other beings.

For Weil, the distinction between paradoxical poles, including the contrary pulls of nature and culture, call for building bridges, not walls. The divide between nature and culture and our paradoxical repulsion and desire for nature cannot be resolved by dividing the realm of the good from that of the earth (as is common with the dogmatic Platonists referenced in the second part of this book). Nor does it call for constructing houses upon our bridges (as is common with rational as opposed to mystical hermeneutic approaches to reading Plato).[28] Weil's notion of houses on bridges points to the tendency in academia to reify a living breathing philosophical process. This often involves a division of thought or theory from action. Weil notes that this is precisely what has been done in the case of popular Platonists' interpretations of Plato: the good is divided from the earth, and thought is divided from action (*IC*, 145). Weil's is a phenomenological rather than a formal ethics and this perhaps indicates certain possible ways of expanding Weil's ethical concerns. For example, Smith uses Max Scheler's work on sympathy and love to explain phenomenological ethics: " 'fellow feeling' can, and often does, grow to include a wider community of significance while not neglecting the importance of the phenomenal experience of the individual case that initially instigates such emotional involvements."[29] Smith notes that such a style of ecological ethics differs radically from the more traditional environmental ethics in that Scheler's ethics is based on love and "Fellow-feelings" as opposed to the "logical (necessary and sufficient) criteria for 'moral considerability' (Goodpaster 325) which are then expanded into moral axiologies – that is, formal systems and hierarchies of values (Smith, *Ethics of Place*)."[30] Smith regards this predominant western ethical style as problematic because such a method sets out to "initially try to define and demarcate those categories of 'beings' that can justifiably be deemed to be morally significant others, from that 'stuff' and those 'things' which logically cannot and are thereby reduced to being, at best, of merely instrumental value to the 'beings' that really (ethically) matter."[31] To place Smith's account in the context of this book, the traditional formal ethical model has often been a matter of reinforcing a divide between nature ("things" without ethical significance) and culture ("beings" with ethical significance) that is not helpful to truly recognizing and living the phenomenological and material interconnection of ethical demands. Weil's system of thought has a clear ethical duty towards other humans and a recognition of one's position with and amongst others, and it similarly defies the reification and strict moral guidelines of traditional ethical structures. As such, it also makes possible this type of more expansive ecological ethics. In part this is foreshadowed in her notions of attaining a balance between force and necessity, but it is also related directly to her notion of decreation which, I believe, can be used to think and respond to the immense ethical concerns raised by the Anthropocene.

Weil's philosophy takes the care of the other to be the highest virtue and indeed the most relevant virtue for both the other and the self. Her ethics is the practice and expression of love that is never perfect. It is not possible to reach the Good, but we can *attempt* to reach the Good, and that is enough despite the inevitable failings along the way. This phenomenon of cultivating ethical theories and actions even when they "fail" in practice is considered by Keller when she notes that the movements inspired by Mahatma Gandhi and Martin Luther King, Jr. were not straightforward successes when we consider "the assassinations, betrayals, disappointments, the backlashing generations" and yet "they came closer to realizing their own hopes than have most political events of assemblage for justice."[32] Gandhi and King's meditative practices of spiritual silence and attention, which are a major inspiration for social justice movements to this day, bear a striking resemblance to Weil's ethical work. For Weil, it is a matter of continually going through the process of decreation, of cultivating the faculties and the existential barriers to those faculties through attention, finding balance between them, and crossing the bridge of that balance into action. In setting out a Weilian-inspired decreative ecological ethics, I will attempt to theorize this movement with a comparison between my own mapping out of Weil set out in the first part of this book and the ecological ethics and concepts associated with the movement and philosophies referred to as deep ecology.

Deep Ecology and Decreation

The significant overlaps between key concepts in deep ecology and Weil's decreation include ecological forms of attention, as well as the expansion of a non-egoistic ecological self alongside the possibility of developing a Weilian non-anthropomorphizing recognition of non-human things and beings. I see this as a helpful way to interrogate the possibility of decreation in the ecological sphere, because many of the critiques and questions that may arise in regard to this project have already been grappled with in the realm of deep ecology.

Originating with the philosopher Arne Naess, deep ecology is an environmental philosophy involving a process of personal self-realization through the death of the ego-driven small-self and recognition of the *ecological self* as a part of the larger ecological community. Naess explains:

> Ecology (and especially human ecology) teaches us daily more about certain kinds of dependencies. The manifestations of the capacity of sympathy and symbiosis teach us that there is a vast variety of ways of living together without destroying others' potentials of realization. Maximal realization of potentials implies the utilization of the existing diversity of life-forms and capacities.[33]

Identification with other beings in particular encounters cultivates a recognition of the ecological self through empathy towards those beings: the ecological self is realized by identifying with other beings, which happens when we cultivate empathy.[34] This is where the "notion of a being's interest furnishes a *bridge* from self-love to self-realization" in an ecological sense.[35] In this way, care for the other is not self-sacrifice, which Naess regards as a potentially treacherous ground for conservation, but a realization of a fundamental expansive ecological interconnectivity. It is only through going outside the small-self that the self can be expanded into an ecological self through empathy for the other. Warwick Fox goes further, drawing on James Lovelock and Lynn Margulis' notion of Gaia, explaining that this process of self-realization is also party to the realization of Gaia, or the earth: in a somewhat Hegelian sense, it is the process of everything realizing itself.[36] But George Sessions notes that Naess thinks through self-realization predominately from a human perspective rather than a Gaia-centered perspective.[37] Naess argues that self-realization springs from the "ecological self," which is a distinctly human realization of the self. There are two potential avenues of self-realization then, and in Weilian terms, the one from Naess is akin to the decreation of the self at the site of the other. Fox offers the addition of a higher power in that action, as the Weilian use of God experiencing nature through the human as the blind man walks with his stick. As such, both levels of self-realization are present in Weilian decreation on her supernatural and natural levels of experience.

There is a sense of fulfillment in the expansion of the small egocentric self to the more expansive ecological self. It is a fulfillment of self-realization that implies a pleasure. However, Naess distinguishes self-realization from pleasure or happiness, which he interprets as typically human experiences, whereas self-realization is something that can be attributed to other beings such as the male praying mantis who mates and is eaten by its mating partner in order to supply the future offspring with enough sustenance to survive. This act is unlikely to be one of "happiness" for the male mantis (although he does not deny this is possible), but instead Naess conceives of it as an expansive act of self-realization.[38] He maintains a hierarchical place for the human as the central figure of self-realization: "The rich reality is getting even richer through our specific human endowments; we are the first kind of living beings we know of which have the potentialities of living in community with all other living beings."[39] As such, Naess does not share the Gaia-centric interpretation of self-realization that Sessions and Fox promote.

In Sessions' view, humans are not necessary for self-realization. For Sessions (and Fox) all beings are a part of an unfolding process of self-realization: "everything – the system – is realizing itself … if humans were not here anymore, there would still be Self-realization."[40] Thus, the Gaia-centric interpretation is a Hegelian-like system, in which humans can but need not take part, differentiating it from Hegel in this important way because for Hegel, self-recognition by the Absolute requires human intellect, namely, that of Hegel himself. Fox notes that this process of self-realization is interpreted by the majority of deep ecologists, Naess included, using the Spinozist cosmological and psychological path towards self-realization in which all entities are a part of a single process of becoming.[41] Fox notes that the Spinozist interpretation of self-realization argues that "all entities are part of a single, unfolding process … when people attain a deep seated realization of this understanding they will scarcely be able to refrain from identifying with 'other' entities."[42] While self-realization is the supreme norm of deep ecology, there is no end point to "realization"; an eternal or permanent self is not postulated in this process. Rather, it is a process initiated by one's experiences in nature wherein the subject, object, and the medium all come together in an event. The fully realized self is a symbol of identification with an absolute maximum range of beings.[43] This concept is Spinozist in a number of ways: it affirms a continual process of becoming, understands the cosmological as reliant and interconnected to the ethical, and traces the "self-realization" of an "ecological self" back to the Spinozist notion of "self-interest" or seeking one's own profit.[44]

The ecological self has strong similarities to the Weilian decreated self and as such, merits further unpacking. As Weil links decreation to Buddhist detachment, Naess links the ecological self to Eastern religious practices, most notably Zen Buddhism. Fox notes that Zen practice emphasizes a transformation of personality and a dispersion of personal attachments rather than a cultivation of specific moral precepts: "Rather than merely imposing precepts from the outside, their observance emerges from within as a by-product of the change in consciousness zazen [Zen Buddhist meditation] can bring."[45] Naess similarly asserts the supremacy of phenomenologically experienced environmental ontology and realism over fixed codes of an environmental ethics: "If reality is like it is experienced by the ecological self, our behaviour naturally and beautifully follows norms of strict environmental ethics."[46] To see a maximal number of others as the "I" implies acting ethically towards oneself. For example, if the lake that I live beside is an extension of who I am, then it follows that I will be inclined to treat it in an ethical way. This ontological understanding of the lake as an extension of the I implies the ethical treatment of the lake. For Fox, the key tenet of deep ecology is precisely this "transpersonal" approach,[47] which unites the thinkers both "in opposition to approaches that issue in moral 'oughts'" and "in subscription to the approach that I have referred to as transpersonal ecology (i.e., the this-worldly realization of as expansive a sense of self as possible)."[48] From the seat of the ecological self, there is no difference between loving the self and loving the other. To care for the lake is to care for myself.

Bill Devall notes that the distinction between rule-based environmental ethics and ecological consciousness "has practical implications. It is not just a disagreement among some philosophers … [The environmental ethics approach] fails to touch the core of the self."[49] For Devall, expanding that which we identify with, and so expanding our sense of self, creates a feeling of joy and allows for a recognition of "ethical" action. There are still conflicts in such a model, but a "deeper perception of reality and deeper broader perception of self" allows for less conflict at the level of action.[50] Naess argues that this ecological self is cultivated through an expansion of one's own identity: "one's own self is no longer adequately delimited by the personal ego or the organism. One experiences oneself to be a genuine part of all life."[51] Emphasizing an ethics derived from self-cultivation, Aldo Leopold, a pioneer in modern environmental ethics who pre-dates deep ecology, notes: "No important change in ethics was ever accomplished without an internal change in our intellectual emphasis, loyalties, affections, and convictions."[52] This cultivation of the ecological self involves an emphasis on each being's own interest in its self-realization. Indeed, the capacity for self-realization is the highest value in acting ethically. Freya Matthews explains: "the primitive form of value is the interest-in-self-realization that is embodied in a self, or a self-maintaining system."[53]

The cultivation of an ecological self demands a shift from mere thought to action. This means that deep ecology is a favored school of thought for activists. John Seed, who identifies as an activist as well as a deep ecologist, explains: "ecological ideas are not enough, we need an ecological identity an ecosophic self. Ideas only engage one part of our mind in cognition. We also need ecological feelings and actions as well as ideas to nurture a maturing ecological identity in a place."[54] Alongside Joanna Macy, Seed began holding workshops called "the Council of All Beings" that seek to experience the ecological self in ways beyond rational thought.

Deep ecology's notion of an ecological self is the main connection I would like to make to Weil's decreation. As I have set out in Chapter 5, decreation similarly involves the death of the small-ego-driven or individuated sense of self, and it gives birth to a broader, relational sense of self. An openness to mystery and difference is at the heart of this decreative ethical process for Weil. This brings out a major distinction between decreation and deep ecology. While the decreative process places ethics as *a priori* and is driven by reconciling Weil's experience of the divine with her experience of immense suffering in the world, deep ecology posits ontology and our being in the world as what leads to the ethical expansion from ego-I to eco-I. Deep ecology's self-interest is part of the "engine" of self-realization; in pursuing my self-interest, I end up generating a more all-encompassing good. In the process of self-realization, one becomes ethical. This is distinct from Weil in that decreation is a response to the phenomenological experience of the other's suffering. As this suggests, there are serious differences between deep ecology and decreation, as would be expected given their very different foci, even though the formulation of an expansive self-understanding expressed in ethical action is vitally important to both. I will now briefly elaborate on some of these major differences so as not to downplay them and to avoid conflating the two approaches. Paying attention to these distinctions also brings out some of the philosophical nuances of both and helps us to recognize how Weil ultimately provides a rather different path into ecological thought and action.

The Way of Joy and the Way of Affliction

One major point of divergence between Naess and Weil is the position of joy as opposed to affliction. Naess links self-realization with a sense of joy. He calls upon a Spinozist inner relation between joy and the increased *power of realization* in opposition to sorrow with the decrease in the power of realization. In this interpretation, joy is an inherent part of the process of realization.[55] For Weil, however, decreation happens predominately through the destructive force of affliction. It is a painful rending that destroys one's sense of the small-self. It requires dropping into an absolute void, floundering without meaning and only then, through careful attention to the contradictions of existence, rising into a higher void. Even in the elation of the higher void or the beauty of the world, joy would not be the word to describe what Weil posits. There is a unity, a connection to the Good, but ultimately this process remains imperfect. Just as Socrates is the wisest philosopher in that he knows he knows nothing, a glimpse of the Good serves as recognition that one is distinctly not the Good and at best is on the way towards the Good. Following the logic of Diotima, those seeking the Good necessarily lack the Good.

While she finds affliction to be the most efficient, it is not the only way to access the Good or the process of decreation. Indeed, love of the beauty of the world is incredibly important to this recognition, as is the love of the neighbor, religious practices, and friendship (*WG*, 83–142). Similarly, Naess' recognition of joy in nature only becomes necessary through experiences of the afflictions imposed by modern forms of anti-ecological development. The actions that emerged from and gave rise to deep ecology have also been associated with opposing these developments. Fox explains that Naess

> gave up his professorship and emerged from academic isolation in order to be freer to participate in the multitude of popular campaigns for ecology and social change. His fearless action has added weight to these campaigns, and the well-known picture of the internationally renowned professor calmly being carried away by the police from the protest camp at Mardøla has certainly given

many good citizens a new understanding that activists are not only "hysterical extremists."[56]

A consequence of and impetus for both Naess' deep ecology and Weil's decreation is a drive towards action and justice as inseparable from the life of the mind. While Weil privileges affliction, Naess emphasizes joy. But both thinkers maintain the necessity for both affliction and joy and the ultimate goal of translating thought into ethical action.

Attention

The ethical practice of attention is of the utmost import to both thinkers. Between her early Marxist materialist work and her later religious inclination, the direction of Weil's attention shifts. For the early Weil, attention moves from "the general to the particular, from the abstract to the concrete" (*LP*, 59). A notable exception to this rule is art, which Weil takes to be uniquely particular. *Notre Dame de Paris* is not a church first, it is *Notre Dame* first. This priority of the general over the particular seems to shift after her religious recalibration.[57] After this shift, it is the act of particular individual decreation that allows one to slough off the egocentric false-I and recognize the decreated larger sense of the I. What one discovers at the core of the particular is precisely the general or the impersonal desire for the Good. Thus, the particular recognizes in itself the general as that which makes it morally worthwhile. So, while there is a shift in the uncovering of the general, it is still the general and not the particular that makes a particular thing morally important. Furthermore, in that we are finite beings, decreation happens to us whether we choose it or not. The fact is that all forms of particular meaning will inevitably be torn away from us during this earthly existence. What we *can* choose is the cultivation of a decreative attitude within our lifetime.

Naess approaches attention from the opposite end, "thinking of the realization of the potentials of particular living beings" first and foremost.[58] It is the particular that is morally worthwhile and not the general. But while they approach the practice of attention from opposite ends, they each gesture to a respect for the other approach and a recognition of its importance. Neither Weil nor Naess denigrates the particular or the general, even if they do take one or the other as their starting point. A bumble bee is both a member of the *Bombus* genus, comprised of 250 species,[59] and it is also *this* bumble bee, who has chosen to pollinate this particular flower in the window box outside my office as I write. I think that it is unwise to conceptualize the particular in a hierarchical opposition to the general. Instead, they are in a circular process of co-creation.

Expanding the Self

Following Spinoza, Naess argues that self-interest is identical with virtue, but only if we take the concept of self in the broader sense.[60] In Spinoza's ethics this concept arises from the notion of *conatus* or striving. This striving is the mode of all particular things including the human mind: "The mind, both in so far as it has clear and distinct ideas, and also in so far as it has confused ideas, endeavours to persist in its being for an indefinite period, and of this endeavour it is conscious."[61] This leads Spinoza to the ethical stance that to strive for what is good for a person means that this person is virtuous:

> The more every man endeavours, and is able to seek what is useful to him – in other words, to preserve his own being – the more is he endowed with virtue; on the contrary, in proportion as a man neglects to seek what is useful to him, that is, to preserve his own being, he is wanting in power.[62]

Spinoza seeks to prove this statement with the following argument:

> Virtue is human power, which is defined solely by man's essence (IV. Def. viii.), that is, which is defined solely by the endeavour made by man to persist in his own being. Wherefore, the more a man endeavours, and is able to preserve his own being, the more is he endowed with virtue, and, consequently (III. iv. and vi.), in so far as a man neglects to preserve his own being, he is wanting in power. Q.E.D.[63]

According to Spinoza, the essence of human beings is this striving for self-preservation and as such, this *conatus* is a virtuous power. Naess adapts this notion of the striving within all things, applying it to a larger sense of the self.

For Naess, ecological ethics must overcome the fallacious ethical distinction of self-love as selfish and love of the other as virtuous. Instead, the self must be broadened so that all love is self-love, and all self-love is love of others. In this way, ethical ecological action is self-serving but also serves the larger more-than-human community. The two are not mutually exclusive and indeed, they are firmly entangled: "We may be said to be in, of and for nature from our very beginning. Society and human relations are important, but our self is richer in our constitutive relations."[64]

Naess' morally charged notion of self-love as an expansion of the self is by no means held across the board for deep ecologists. Through this broad self-interest, Naess suggests that one *wants* to protect all beings as extensions of the ecological self, whereas Fox argues that one *ought* to protect all beings regardless of one's own psychological relation to those beings and as such, Fox maintains ethical duty beyond one's own human desire.[65] J. Baird Callicott makes a similar argument about the expansive relationality of an ecological self based on the Buddhist sentiment of the world as one's own body.[66] Smith argues that such a claim has no ethical fecundity: "The fact that we are at one with nature at this level gives us no ethical guidance at all, for so too are murderers, logging companies, and industrialists."[67] Similar to Weilian ethics, the eco-self is a recognition of our obligation to others rather than a response within the rights-based crime and punishment model. It establishes the dignity of things and beings rather than seeking to divide them into those that are innocent and those that are guilty. The extent to which the ecological self creates or determines any necessary decision about what constitutes specific ethical action is by no means settled among deep ecologists and their critics. For both Naess and Weil, the expansion of the small-self to a larger relational self implies an expansion of empathy and a recognition of this broader self as always already intwined with others. Ethical action springs from the recognition of interconnection and worth within each thing one encounters. It is not merely the fact of interconnection that generates ethics, but consciousness of that interconnection. Following Weil and Naess, this consciousness has an affective as well as cognitive dimension, realized in action.

In this ethical structure put forward in Weil and Naess, one is cultivating virtue in one's self and in so doing simultaneously caring for the other prior to the self. Naess contends that "The higher the level of realization of the potentials of a living being, the greater the dependence of further increase in level upon the increase of the level of other living beings."[68] By this, he suggests that our self-realization is an increased awareness of our vulnerability to and reliance upon others. A narrow sense of individual autonomy is a sign of being stunted in one's own personal realization. On the other hand, increased interdependence, and the promotion of self-realization in others, is a sign of advanced self-realization. Naess compares this to Mahayana Buddhism in which full enlightenment is not reached by one until it is reached by all sentient beings.[69]

Non-anthropomorphizing Recognition

Gisli Palsson and Heather Swanson note that the Anthropocene challenges us to reimagine "how to craft a planetary that itself brings difference to the fore."[70] Weil's decreation offers an ethical, practical mode of bringing difference to the fore without attempting to totalize it. But how do we recognize the unique difference of the more-than-human other without anthropomorphizing and totalizing that difference? Totalizing the other happens so easily and often without intention. It can arise from caring and loving and not necessarily from any sort of ill-intent. For example, when reflecting on the care-filled act of bee keeping, Farrier notes the overlapping of love and care with "the violence of managing others."[71] To express care sometimes means to control, to dismiss the mystery of difference.

Naess suggests that we can avoid anthropomorphizing more-than-human difference by considering every living being's process of self-realization as a realizing of potentialities that are inherent in a being and not merely projected onto them. So while my empathy for another being may set in motion a recognition of the other as useful for my own self-interested pursuits, ultimately, the realization of that being's inherent potential is an objective point of study unique to the plant, animal, or thing and not contingent upon human empathy. In this process, we realize ways of living together with all manner of beings without destroying the potentiality of those beings.[72] Fox points out that all living beings can be said to embody certain kinds of interests and any kind of being that has interests deserves moral appreciation.[73] This is regardless of a being's ability to self-reflect.

In responding to some of deep ecology's critics, Fox defines anthropocentrism as "unwarranted differential treatment of other entities on the basis of the extent to which they are considered to be human-like."[74] He then notes that while deep ecology focuses on the human capacity to self-realize, via the ecological self, it is not anthropocentric. Self-realization focuses on the human ecological self as a means of reimagining the human as an integral part of the more-than-human world. This focus on human methods of overcoming the small anthropocentric self should not be taken to be anthropocentric but rather a practical way for humans to take part in this process. It is the place where humans, situated in human being, have to start. In contrast to anthropocentrism, the human is where this process starts, but it is not where it ends.

Deep ecology is not just charged with being anthropocentric. Smith notes that from the other end of the spectrum, deep ecology is charged with being anti-human. For Smith, the purpose of deep ecology taking a "position against the unquestioned assumption of human sovereignty is not to disparage or denigrate human beings; it is to take sides in an argument about the ethical importance of more than just humans."[75] To disavow human supremacy is not to disavow the human. This is an important distinction. While not a deep ecologist herself, Haraway also suggests that non-anthropomorphizing recognition can happen through making kin with more-than-human others and in so doing recognizing oneself as a part of an ecological community beyond the boundaries of one's own skin.[76]

The hermeneutic structure that emerges from this eco-philosophy is one that privileges the *interesse*, or in Platonic Weilian terms the *metaxu*. For example, Naess notes that the joy I may feel looking at a beautiful tree is not solely in me as the individual viewing the tree, nor is it in the tree, but rather "the joy is a feature of the indivisible, concrete unit of subject, object, and medium."[77] Naess simply states that all beings, like human beings, are complex and we see only parts of them at once. The maple tree in the deep snow of winter is radically physically different than it is in the heat of July, but then again, Naess argues that we see a totally unique facet of the tree in the evening as opposed to just eight hours earlier that morning. All beings are uniquely multifaceted and reveal only parts of themselves to one human's limited perception at any given time and space. As described above, Fox refers to this mode of interconnection as a "transpersonal approach to ecology."[78]

Eco-critic Timothy Luke reads the concept of self-realization as an ideological anthropocentric self-serving personification of the self into nature.[79] Luke argues that by ignoring sociopolitical dimensions of ecological ethics, deep ecologists slip into ideology and project themselves onto nature for the sake of a self-serving ethical transcendence. Also pushing back against deep ecology's resistance to the sociopolitical, Smith argues that social political theories must be taken seriously for how they engage with ecologies, and this is absolutely the spirit of my use of Weilian- inspired ecological ethics. The marriage of deep ecology and social theory that Smith suggests is not necessarily an easy fit, and despite the potential inspiration and use value of such a union, Smith notes:

> Yet deep ecology, as a field, might find engagement with at least some of the various and ever changing forms of social and cultural theory both inspiring and useful, and this despite the serious ecological problems these theories might have when taken as a whole and/or in their usual academic habitats, and even their explicit antipathy of some of their proponents towards deep ecology. I do not want to underestimate the difficulties here.[80]

These difficulties present a series of challenges but are not insurmountable. Similarly, Weil's social-justice-oriented materialism strikes the reader as strangely paired with her religious turn into decreative theology. However, these two aspects of her thought serve to inform one another if one avoids the collective dogmas of either doctrine, just as social theory and deep ecology can resonate if one does not take Naess' tenets as absolutely fixed nor one's social theory of choice (be it Marxist, feminist, etc.) as the absolute or sole purveyor of truth. Integrating ecology and politics requires delicate and nuanced philosophical attention: a practice of attention that is able to sustain tensions and even contradictions in the "between."

For our purposes, perhaps the most notable similarities between decreation and deep ecology are the concepts of attention and the expansion of the self that dislodges the I's illusion of ecological centrality. As such, deep ecology is an apt conceptual vehicle for envisioning how Weil's ethics could be expanded to the more-than-human world. Ultimately, some of the major differences between the eco-self and decreation lead me to why Weil's ethics is preferable in thinking through human duties and obligations in the Anthropocene. Weil's emphasis on affliction and suffering, which is merely tempered by the joy that grounds the eco-self, allows for a realism that does not romanticize or potentially lose oneself in "nature" or the Good. There is not a space for self-serving transcendence in Weil, which allows for the grim realities of anthropogenic climate change to be felt and intellectually engaged. Furthermore, because Weil begins this process from ethical obligation in the face of the other, rather than deep ecology's foundation of ontology and self-interest through self-realization, ethical care of the other and the larger self as relational is foundational in decreative ethics. In deep ecology, ethics is a secondary mode, a by-product of the transformation of self from ego to eco. This difference keeps ethics at the center of Weilian thought in a pivotal way. Like deep ecology, the Weilian-inspired ecological ethics this book will propose seeks to balance having a special and unique place for human dignity, while respecting and recognizing the dignity of the more-than-human world. This recognition arises from attending to the non-anthropomorphizing interests of that world and one's interconnection to other beings and things.

Notes

1 Paul Crutzen, "Geology of Mankind," *Nature* 415 (2002), 23. https://doi.org/10.1038/415023a.

2 Cristophe Bonneuil and Jean-Baptiste Fressoz, *The Shock of the Anthropocene* (London: Verso, 2015), 17.

3 Mick Smith, *Against Ecological Sovereignty: Ethics, Biopolitics, and Saving the Natural World* (Minneapolis, University of Minnesota Press, 2011).
4 Clare Colebrook, *Anthropocene Feminism*, ed. Richard Grusin (Minneapolis: University of Minnesota Press, 2017), xiii.
5 Moore argues: "The Capitalocene signifies capitalism as a way of organizing nature – as a multispecies, situated, capitalist world-ecology." See Jason W. Moore, *Anthropocene or Capitalocene? Nature, History, and the Crisis of Capitalism* (Oakland: PM Press, 2016), 6.
6 Donna J. Haraway, *Staying with the Trouble: Making Kin in the Chthulucene* (Durham: Duke University Press, 2016).
7 Anna Lowenhaupt Tsing, *The Mushroom at the End of the World: On the Possibility of Life in Capitalist Ruins* (Princeton: Princeton University Press, 2015), 19.
8 David Farrier, *Anthropocene Poetics: Deep Time, Sacrifice Zones, and Extinction* (Minneapolis: University of Minnesota Press, 2019), 9.
9 Catherine Keller, *Political Theology of the Earth: Our Planetary Emergency and the Struggle for a New Public* (New York: Columbia University Press, 2018), 5.
10 William Cronon, "The Trouble with Wilderness; or, Getting Back to the Wrong Nature," in *Uncommon Ground: Rethinking the Human Place in Nature*, ed. William Cronon (New York: W.W. Norton, 1995), 69–90. For a response, see Donald M. Waller, "Getting Back to the Right Nature: A Reply to Cronon's 'The Trouble with Wilderness,'" in *The Great New Wilderness Debate*, eds. J. Baird Callicott and Michael P. Nelson (Atlanta: University of Georgia Press, 1998), 540–567.
11 Henrik Ernstson and Erik Swyngedouw, "Framing: Rupturing the Anthro-obscene! The Political Promises of Planetary and Uneven Urban Ecologies," at *Conference Teater Reflex*, September 16–19, 2015, organized by KTH Environmental Humanities Laboratory, Stockholm (online: www.researchgate.net/profile/Henrik_Ernstson/publication/280738621_Framing_the_Meeting_Rupturing_the_Anthro-obscene_The_Political_Promises_of_Planetary_Uneven_Urban_Ecologies/links/55c4ab2608aebc967df374c4.pdf, accessed October 21, 2016).
12 Bill McKibben, *The End of Nature* (New York: Random House, 1989).
13 Kristin Asdal, "The Problematic Nature of Nature: The Post-constructivist Challenge to Environmental History," *History and Theory: Special Issue on Environment and History* 42.4 (2003), 61.
14 Ibid.
15 Tsing, *Mushroom*, 218.
16 Farrier, 51–52.
17 Ibid., 84.
18 Ernstson and Swyngedouw.
19 "nature, n.". 11, *Oxford English Dictionary* Online. March 2021. Oxford University Press (www-oed-com.proxy.queensu.ca/view/Entry/125353?result=1andrskey=gPwZ4Vand, accessed April 13, 2021).
20 "nature, n.". 5a, b, *Oxford English Dictionary* Online. March 2021. Oxford University Press (www-oed-com.proxy.queensu.ca/view/Entry/125353?result=1andrskey=gPwZ4Vand, accessed April 13, 2021).
21 Bill McKibben, "The Death of Nature," in *The Norton Book of Nature Writing*, ed. Robert Finch and John Elder (New York: W.W. Norton, 2002), 1121.
22 Michael Smith, "Deep Ecology: What Is Said and (to Be) Done?" *The Trumpeter* 30.2 (2014), 146.
23 Farrier, 12.

24 Smith, "Deep Ecology," 146.
25 Karen Barad, *Meeting the Universe Halfway: Quantum Physics and the Entanglement of Matter and Meaning* (Durham: Duke University Press, 2007), 30. Quoted in Farrier, 70.
26 Robert Pogue Harrison, *Gardens: An Essay on the Human Condition* (Chicago: University of Chicago Press, 2008), 165.
27 Ibid.
28 For Weil's suggestion to read Plato as a mystic, see Weil, *Seventy Letters*, 86.
29 Mick Smith, "Dis(appearance): Earth, Ethics and Apparently (In)Significant Others," *Australian Humanities Review* 50 (May 2011), 28.
30 Ibid., 29.
31 Ibid.
32 Keller, *Political Theology*, 122.
33 Arne Naess, "Self-Realization in Mixed Communities of Humans, Bears, Sheep, and Wolves," *Inquiry* 22 (1979), 231–241.
34 Arne Naess, "Self-realization: An Ecological Approach to Being in the World," *The Trumpeter: Voices of the Canadian Eco-philosophy Network* 4.3 (1987), 35.
35 Ibid., 36. My emphasis.
36 Warwick Fox, "On the Interpretation of Naess' Central Term 'Self-realization,'" *The Trumpeter* 7.2 (1990), 98.
37 Naess, "Self-realization: An Ecological Approach," 42.
38 Ibid.
39 Ibid.
40 Quoted in Fox, "On the Interpretation of Naess' Central Term," 99–100.
41 Fox, "On the Interpretation of Naess' Central Term," 99. Fox notes that the other option suggested by some deep ecologists is an axiological perspective in which self-realization has an intrinsic value.
42 Ibid.
43 Ibid.
44 Naess, "Self-realization: An Ecological Approach," 37.
45 Daniel Goleman, *The Varieties of the Meditative Experience* (London: Rider, 1978), 95. Quoted in Warwick Fox, *Toward a Transpersonal Ecology: Developing New Foundations for Environmentalism* (Boston: Shambhala, 1990), 218.
46 Naess, "Self-realization: An Ecological Approach," 40, 36. Quoted in Fox, *Toward a Transpersonal Ecology*, 218.
47 Fox lists the deep ecologists whom he interprets as sharing this belief in a transpersonal ecological self: George Sessions, Bill Devall, Andrew McLaughlin, Alan Drengson, Michael Zimmerman, Lorne Leslie Neil Evernden, John Livingston, John Rodman, and Joanna Macy. Fox, *Toward a Transpersonal Ecology*, 226–232.
48 Fox, *Toward a Transpersonal Ecology,* 224.
49 Bill Devall, "Issues in Contemporary Ecophilosophy," paper presented to the *Ecology and Society Conference*, University of Wisconsin at Waukesha, April 7, 1984, p. 8. Quoted in Fox, *Toward a Transpersonal Ecology*, 226.
50 Bill Devall, *Simple in Means, Rich in Ends: Practicing Deep Ecology* (Utah: Gibbs Smith, 1988), 42–44. Quoted in Fox, *Toward a Transpersonal Ecology*, 226.
51 Arne Naess, "Self-realization: An Ecological Approach," 39–40. Quoted in Fox, *Toward a Transpersonal Ecology,* 230.
52 Aldo Leopold, "Land Ethic," in *A Sand County Almanac: With Other Essays on Conservation From Round River*, ed. Charles Walsh Schwartz (New York: Oxford University Press, 1966), 225.

53 Freya Matthews, *The Ecological Self* (Abingdon: Taylor and Francis Group, 1991), 82.
54 John Seed, "The Ecological Self," *The Trumpeter* 22.2 (2006), 98. "Ecosophic" is a term denoting ecological philosophy.
55 Naess, "Self-realization in Mixed Communities," 233.
56 Warwick Fox, "Arne Naess: A Biographical Sketch," *The Trumpeter* 9.2 (1992). Online.
57 For more on this, see my first chapter and my treatment of Weil's mystical thinking.
58 Naess, "Self-realization in Mixed Communities," 234.
59 Paul H. Williams, "An Annotated Checklist of Bumble Bees with an Analysis of Patterns of Description," *Bulletin of the Natural History Museum, Entomology Series* 67 (1998), 79–152.
60 Naess "Self-realization: An Ecological Approach to Being in the World," 37–38.
61 Benedict de Spinoza, *The Ethics*, trans. R.H.M. Elwes (online: Project Gutenberg, 2003) Book III, Proposition IX.
62 Ibid., Book IV, Proposition XX.
63 Ibid., Book IV, Proposition XX, Proof.
64 Naess, "Self-realization: An Ecological Approach to Being in the World," 35.
65 Fox, "On the Interpretation of Naess' Central Term," 99.
66 J. Baird Callicott, "Intrinsic Value, Quantum Theory, and Environmental Ethics," *Environmental Ethics* 7.3 (1985), 274.
67 See ibid. and Mick Smith, *An Ethics of Place: Radical Ecology, Postmodernity, and Social Theory* (New York: SUNY Press, 2001), 35.
68 Naess, "Self-realization in Mixed Communities," 236.
69 Ibid., 39.
70 Gisli Palsson and Heather A Swanson, "Down to Earth: Geosocialities and Geopolitics," *Environmental Humanities* 8.2 (2016), 155. Quoted in Farrier, 17.
71 Farrier, 107.
72 Naess, "Self-realization in Mixed Communities," 232. Farrier also notes the difficulty in empathizing with beings that do not have "a center," as it were: "Jellyfish lack a center upon which human attention and care can gain purchase. They do not have eyes or anything resembling a face; they don't even have brains, possessing instead a diffusely arranged 'nerve net.' No center implies no self." Farrier, 95.
73 Fox, "On the Interpretation of Naess' Central Term," 99.
74 Ibid., 101.
75 Smith, "Deep Ecology," 148.
76 See Donna J. Haraway, *Staying with the Trouble* (Durham: Duke University Press, 2016). Also see Adele E. Clarke and Donna Haraway, eds., *Making Kin Not Population* (Chicago: University of Chicago Press, 2018).
77 Naess "Self-realization: An Ecological Approach to Being in the World," 41.
78 Fox, "On the Interpretation of Naess' Central Term," 101.
79 Timothy W. Luke, *Ecocritique: Contesting the Politics of Nature, Economy, and Culture* (Minneapolis: University of Minnesota Press, 1997), 16–18.
80 Smith, "Deep Ecology," 149.

9 A Weilian-Inspired Ecological Ethics

Weil's work as a whole, and her ethical system in particular, is not necessarily a natural fit for an ecological ethics. While I maintain that a strong ecological ethics can arise from her work, it can only do so with careful attention, interpretation, and even gentle shifts in the direction of her work. With this in mind, I will expand on the possibility of extending ethical consideration to more-than-human life in Weil's thought and then turn to the notions of humanism, God, and purity in Weil. All three of these aspects of Weil may be seen as problematic in light of my proposed ethics as a matter of recognizing the impersonal connectivity underlying all being and in all beings. In each case, I will appeal to my non-dualistic Weilian reading of Plato as a ground for understanding how these concepts can still work within this ecological ethics. Additionally, Weil's work will be supplemented and shifted in engagements with deep ecology and myriad other philosophers to the point that we must consider this to be an ecological ethics inspired by Weil but certainly not one bound to the letter of Weil's writing. Weil's approach may be directly extended into discourses of environmental justice for human benefit – but the goal of this book is to develop a more radical view of Weil as a basis for an ecological ethics with the more-than-just-human environment. With this in mind, I will stop short of asserting an absolute equality between all things. The reasons for this are an ethical pragmatism and the need to weave social political theory into our ecological thought. Because of the staggering predominance of environmental racism, I do not wish to assert that a human life has an equal amount of worth as an animal life or the life of a tree or even that of a waterway. Such an assertion, while it may have good intentions of epistemically elevating the lives of more-than-human beings, also creates an ethical possibility of choosing a tree over a human, which does not amount to sacrificing *any* human, but rather to sacrificing first and foremost the human already closest to being displaced. The humans who will suffer from such a way of thinking are often not from the academic circles trying to save the earth from the thoughtless masses and the insatiable elite, but from the most marginalized and often racialized communities of people. As such, this ethical system will include the more-than-human world but will maintain a strong sense of the human dignity that Weil upholds in her work.

DOI: 10.4324/9781003449621-13

In my Weilian-inspired ecological ethics, the radical other can and must be expanded beyond the other human. As Yoon Sook Cha points out, this is possible in Weil through the renunciation of ecological centrality that decreation entails:

> While we are in the habit of thinking of renunciation as a giving up, in Weil's conception, where it is figured both as an offering and as a wrenching of ecological centrality, renunciation reinscribes the self at the site of its evacuation through a relationship with the other implied by its offering.
>
> (*DEB*, 9)

To wrench ecological centrality from the I means to open oneself up to a world in which the small-I is not the center. Rather, one sees the world as a dynamic series of interconnections of which one is but a part. In this way, the small-I is evacuated and a larger concept of the I as relational is reinstated at the site of the encounter between the I and the other. As we pay attention to our perceptions of the surrounding world, as we decreate, this process does not remove us from the earth but rather opens our eyes to our interconnection with the earth. Weil explains:

> Except the seed die … It has to die in order to liberate the energy it bears within it, so that with this energy new forms may be developed … So we have to die in order to liberate a tied up energy, in order to possess an energy which is free and capable of understanding the true relationship of things.
>
> (*NB*, 179)

The radical transformation of the seed is as the radical transformation of the self in Weil, a metaphor she often calls upon. But here we may consider it beyond a mere metaphor. I take seriously Robin Wall Kimmerer's suggestion in *Braiding Sweetgrass: Indigenous Wisdom, Scientific Knowledge, and the Teachings of Plants* that we humans have a lot to learn from plants and that we must include non-human beings (plants, animals, etc.) as vital and intelligent aspects of our own collective root system.[1] For Kimmerer, plants are teachers if we are capable of attending to their ancient wisdom. Vital to Kimmerer's suggestion is also the suggestion that Indigenous persons' knowledge of plants offers answers to these questions. This tradition, which offers a distinct but equally valid way of engaging in this work of ecological ethics, has undoubtedly been more successful in balancing relations to the more-than-human world. I will refer the reader to Kimmerer as a potential starting point for examining this tradition and note that there is a vast, varied, and rich number of texts to engage. In the spirit of Kimmerer's insights, it makes sense to expand Weil's notion of decreation beyond human beings. Indeed, Kimmerer's recognition of the knowledge we can glean from plants casts light onto a blind spot in the western tradition. As Weil notes, "There is an exponential power contained in seed" (*NB*, 615), and we shall set out to recognize and engage with that power.

Expanding Ethics

Weil is fond of calling upon the Cartesian example of the blind man and his stick: humans are the stick through which God can encounter the material realm. Here we can see human exceptionalism at play in Weil's work, in that humans alone are an extension of God and in a sense, a part of the assemblage of God. While the "I" is never fully lost, the process of decreation through attention makes the human an empty vessel for God to experience the beauty of the world. The "I" steps out of the way and allows for an experience of radical difference, which Weil posits as an experience of God. Consider this experience as tapping into the impersonal desire for the Good at the heart of oneself (which Weil posits at the heart of all human beings), or the morally worthwhile universal within the particular shining through one's being. Weil limits her use of the Cartesian stick to an illustration of this link between human (particular desire) and God (universal Good), but I think it is apropos to extend it further in order to comprehend the possibilities of different beings. The ethical importance of difference and also the interconnection to that which is beyond the "I" can both be brought to light through this Cartesian metaphor. Without mentioning Descartes explicitly, Anna Tsing notes:

> In order to survive, we need help, and help is always the service of another, with or without intent. When I sprain my ankle, a stout stick may help me walk, and I enlist its assistance. I am now an encounter in motion, a woman-and-stick. It is hard for me to imagine any challenge I might face without soliciting the assistance of others, human and non-human. It is unself-conscious privilege that allows us to fantasize – counterfactually – that we each survive alone.[2]

Tsing's quote reminds us that we are also unique players in assemblages with the surrounding world. Insofar as humans are reliant on both earth (natural) and God (supernatural) for sustenance, Weil would not disagree. She notes: "It is only from the light which streams constantly from heaven that a tree can derive the energy to strike its roots deep into the soil. The tree is in fact rooted in the sky."[3] Trees must be nourished by both the soil and the light; likewise, human beings require nourishment in both body (via the earth) and soul (via divine love). But there is an ontological difference for Weil between the tree and the human. The tree's need for sun and earth both remain within the realm of necessity or the natural, while the human body is connected to the natural necessity of existence, but the human soul longs for the Good beyond, which it can only access through the *metaxu* or the momentary mediation between the natural and the supernatural. This distinction is what draws the line of human dignity in Weil and asserts a certain human exceptionalism, which we will examine further at the end of this section. For now, I wish to focus on the spatial interconnection that is necessary in this process: the assemblage of God, human, natural world, is absolutely intertwined. In the moments of *metaxu* in which the small-I is decreated and the divine universal witnesses the beauty of the surrounding earth, the larger sense of the I awakens to an expansion of the I that encompasses the individuated self, the universal desire for the Good, *and* the surrounding ecosystem, of which one's small-self is but one part. There is a spatial interconnection at play between individual, surrounding ecology, desire for the Good, and universal Good. Without the surrounding natural world, there is no connection to the Good.

David Farrier's notion of "deep time" intensifies this interdependence with the suggestion that our quotidian existence is not simply a product of interconnection with other beings in the present, but rather is reliant upon the earth's past far before humans evolved and will continue on far beyond human extinction:

> As we live through it, we encounter the deep past and the deep future in the most ordinary situations, such as through the hundreds of "technofossils" – ballpoint pens, smart phones, plastic bottles, artificial knee joints and heart valves, fiber-optic cables, contact lenses Styrofoam cups, plastic banknotes – that surround many of us every day. But deep time has always been deeply embedded in the present of life on Earth. All life on the planet is here by virtue of a debt owed to the long history in which nonlife shaped the conditions of life to flourish.[4]

The material objects we encounter are never merely in the present. Objects such as these technofossils were forged in the creation of the planet in that they emerged out of previous patterns of creative practices or developments. Furthermore, our technologically derived additions to these technofossils will contribute novel patterns to future fossils. As such, they are rooted deeply in the past and imprinted in the far future. These natural objects form an interconnected part of our human, but also more-than-human assemblages. They are extensions of ourselves, enabling the ways we think and live, but they are also individuated objects irreducible to their purpose for us. They are present before us, they hearken back to long before human life on the planet, and they will continue on with our technologically induced transformations for millennia beyond our own existence through the radical impact humans are having geologically. They construct new possibilities and open up these new possibilities into the future. Our connection to the earth, then, is not just spatial but also temporal in that we are informed by the geological prehistoric past and we are informing the geological future for millennia to come. Our assemblage with God and earth is intertwined through space and time.

This spatiotemporal connection between things is also ethical. Weil elevates the human to the status of a unique harbinger of the divine, given the human ability to pay attention. In order to recognize the ethical layer of this connection, we must recognize humanity as uniquely situated to pay attention. In this work of attention, humans can recognize the attentive intuition paid by the non-human beings with whom we share the earth: the intuition of the flower expressed in its orientation towards the sun, the rabbit's consciousness of the wolf, and the tide's response to the moon. Such intuition is physically embedded within human bodies as well but is covered over by our unique capacity for *inattention.*

Other life forms also change habitat in ways that perhaps put them out of sync with other natural forces. They make these changes in response to environmental pressures and opportunities, such as changes in migration patterns. But humans have proven to be especially proficient at the disregard of the surrounding environment through the cultivation of artificial surroundings. Of course, the divide between that which is "natural" and that which is "artificial" is another point of contention when we consider the diverse materials used in the nests of bowerbirds or the massive ecological disruption caused by beaver dams. This is not meant to cast aspersions on the significant medical, technological, agricultural advances of human beings. Rather, I mean to note the consequences of these advances: namely, the possibility of inattention to the surrounding world and the human-induced imbalance between force as a desire for power and necessity as the material conditions of the earth. This points to Diotima's logic from Plato's *Symposium* in that the human need to cultivate attention reveals the fact of our inattention or our ability to cover over our intuition.

In recognizing our unique capacity for *inattention*, we gain an appreciation for the intuition of other beings and our interconnection with those beings that attention reveals. Decreative self-abnegation is not a loss of self completely because, similarly to Abram's account of perception, "Neither the perceiver nor the perceived, then, is wholly passive in the event of perception."[5] Upon paying attention, I do not absolutely disappear so much as "I find this silent or wordless dance always already going on – this improvised duet between my animal body and the fluid, breathing landscape that it inhabits."[6] What does die in decreation is the idea of an absolutely individuated "I." In the death of this radically separate or self-enclosed "I," difference begins to reveal itself through my attention and an "I" that is conscious of itself as relational emerges at the site where the individuated "I" has dissipated. Neil Evernden describes two possible modes of engagement with the natural world as (1) "making the world immanent in the subject … or … a rejection of relationship through the annexation of all subjectivity to the self. The whole is annexed to the part"; or alternatively, (2) "being filled with wonder at it … an individual who is already in the world, and a situation in which it makes no sense to think of independent parts."[7] Moving into the latter is necessary for a robust environmental ethics according to Evernden, and this is precisely the benefit of an ecological ethics inspired by Weil's decreation when the ethical limits are extended beyond humans.

Tsing rightly notes that "many organisms develop only through interactions with other species."[8] In decreating one's self, one is voluntarily thrown into a situation of radical vulnerability. Decreating does not undo the self completely, but it does shake us out of the chains of stability. It is not in simply thinking vulnerability but in living vulnerability that we can change the way we see other beings and our own social analysis. Decreation allows us to think – and even *become* – the new. It asks us to step into an interconnected vulnerability and bear witness to a number of narratives happening simultaneously.

In the ethical dimension of decreation, I allow for the other person to exist, and in so doing, I mark both my deep connection to that other and the simultaneous mystery of that other being. Like me, the other person also longs for the Good, as this desire is not merely a narrow self-interest but rather the impersonal core of all beings. Given the level of cruelty and suffering in the world, this only makes sense if we conceptualize the desire for the Good as the process of self-realization and recognize that it is often covered over by the petty desires of the egocentric small-I. The desire for the Good is that which Weil sees as the impersonal aspect worth protecting in all humans.

Recall that for deep ecology, each being desires self-realization or the realization of its highest potential, which we can consider to be its highest good. On the other hand, Weil's ethics realizes the Good through one's ethical action at the site of the other person's suffering. The cry of the other person brings about the decreation of the I and reassertion of the self as relational to the other who cries out. In my Weilian-inspired ecological ethics, I suggest that this decreative movement in the face of the other's suffering necessarily begins with and prioritizes the cry of the suffering human, but it expands as our ability to pay attention deepens. Our capacity to expand the self and relate to myriad other humans can lead into our ability to attend, hear, listen, decreate, relate, and act ethically with and for the more-than-human world. One can hear the cry of the factory-farmed animal, listen to the once-arable soil drying and burning, allow for the being of the waterway, relate to the howling winds of the hurricane and act ethically.

Decreating and relating to the more-than-human world is by no means ethically simple or clear because it so often defies human-centric constructs of ethics and it certainly does not take into account rights-based innocence or guilt. Take for example the lion whose sustenance necessitates the death of the gazelle. Recognizing each being's own-most possibility of self-realization (or alternately of being harmed) does not bring about any form of codification through rules. Rather, it is a recognition of the gaps and messiness of lived ethical experience. Another example is the invasive species of reed grass, phragmites, that dominate and destroy wetlands across the great lake's region of North America. In phragmites' self-realization there is a destruction of other beings and environments. If this is to be considered ethically, one must note the difference between self-realization or highest potential (a movement towards the Good via necessity) and self-interest or highest amount of power or growth (a movement of Force). The former must be ethically obliged whenever possible, whereas the latter is a matter of violence and Weilian force that is not conducive to ethics.

Following Naess' account of self-realization, these expansive ecological interests are not the result of our subjective anthropomorphizing of the plant or animal, as if this were necessary for the non-human being to matter ethically. Instead, self-realization is an expression of our attentive openness to the realization of another being's inherent potential as an objective point of study unique to the plant or animal, and yet embedded in a relational context or ethical ecosystem that includes all of us.[9] In this process, we realize ways of living together with all manner of beings without destroying the potentiality of those beings while expanding our own potentiality.[10] Warwick Fox points out that all living beings can be said to embody certain kinds of interests and any kind of being that has interests deserves moral appreciation. This is regardless of a being's ability to self-reflect.[11] Conflicts still occur. This recognition of self-realization does not mean that a flat ethical equality covers over all beings. Rather, it implies an acknowledgment of this striving to flourish in all beings and a recognition that humans have the ability and responsibility to critically take that into account. As our attention is refined and deepened, so too is our responsibility.

While our ethical duty is borne of other beings' ability to self-realize and our own entanglement with other beings, it is also nothing rare or special: We share the bond of self-realization with literally all beings. All beings have an interest in realizing their potential, and this impersonal core of all beings calls for our ethical consideration. We may consider some beings unworthy of this consideration, and perhaps some must be held accountable for ongoing harms committed against others, but this does not exclude them from *a priori* moral consideration. We need not tolerate ongoing harm, but rather, we refuse to evacuate someone or something of ethical worth on the basis of what they have done or failed to do. Unethical actions still have consequences within society, but we recognize the value and dignity of those others as we determine how to hold them ethically accountable for their actions. At their core, other beings' impersonal desire for the good bestows them with a beauty, love, and symmetry that we must ethically acknowledge.[12]

Human Dignity

For Weil, there is an ontological distinction between the desire for the Good or the striving for one's highest potential within humans as opposed to within the rest of nature. Indeed, Weil does not engage with this striving by the more-than-human things and beings of the earth. The human alone strives for a Good beyond the world whereas the animal/plant/thing strives always for the natural good when seeking their own-most potential. The human alone accesses the supernatural through the moments of *metaxu*. This necessitates a difference, if not a human exceptionalism, because while the rest of the natural world can act as *metaxu*, it does not experience *metaxu* in the way of the human. Only humans have this desire for the supernatural Good in a way that transcends the natural, even if it is always in and through the natural that these moments of transcendence occur. Recall that the tree's desire for the light is metaphorically supernatural but in actuality is as natural as its desire for the soil, whereas the human need for the earth is natural but her desire for the Good is supernatural. This contains the Weilian possibility of human dignity and difference.

Rather than the assertion found in Weil that humans alone have access to the "supernatural" God or the Good, I propose following the wisdom of Diotima that perhaps humans are the only ones who are divorced from the Good, as they are the only ones who seek it. Just as the human need to cultivate attention reveals our unique inattention, perhaps our drive towards the Good is unique to humans not because the surrounding world exists only in the "natural" realm, as Weil suggests in her own writing, but because each species, genus, thing in the created world accesses its own unique attentive practices, has its own access to moments of *metaxu*, and its own connections to the supernatural realm. These would look radically distinct from human *metaxu* and undoubtedly would not be as rare or fleeting. They would be rich paths of connection we could attend to in the surrounding world. In such a suggestion, I leave behind a core element of Weil but I maintain the Weilian division of "natural" and "supernatural," I continue to assert Weil's non-pantheistic encounters with the divine, and I uphold human difference in so far as the nature of our desire and striving towards the Good or God. I hypothesize that our ethical striving reveals our uniquely human distance from the Good and suggest that the other non-human beings and things of the earth may, in their own way, contain and practice the ethical knowledge and attention we seek. Rather than merely "natural," what if they have achieved a just balance of their own? Recognizing this possibility seems to me to be an outcome of attention in the Anthropocene. The cost of this assertion is that such an ethics is no longer Weilian in that it eschews her fundamental human exceptionalism when it comes to accessing the "supernatural" or the Divine Good. But I maintain Weil's fierce commitment to human dignity and note that this dignity remains unique and morally of the utmost import in that humans alone strive for the Good in our specific vulnerable and precarious way.

As we noted in deep ecology, Naess argues for a human-focused realization of the ecological self that is not about centering the human over and against nature but about recognizing that as humans, we will always come to our understanding of the world through our own unique human phenomenological perspective. This is simply an embodiment of Kant's Copernican revolution, which realizes that as humans we experience the world through our unique human sensual apparatus and as such, the world we experience is uniquely human. This does not mean we cannot be enriched by and engage with (indeed learn and grow with and from) other beings, but rather that humans have a specific dignity founded in our shared perspective. As such, it is pragmatic to consider what that perspective entails, and how it allows us to recognize our interconnection and ethical duty to the surrounding world. Recognizing human difference gives us the opportunity to consider the alternative perceptions and perspectives of non-human others. For Weil, in extending the I in an ethical bind to another person, we recognize the dignity of humans in the recognition that they too desire a supernatural Good. But the natural world's desire does not go beyond a natural good for Weil. And even if we follow my hypothesis that the surrounding world and its more-than-human things and beings have their own unique access to the Good, theirs is not the same structure as human desire and human connection to the Divine Good through *metaxu* and ethical actions. In this sense, the ethical bind to non-humans is weaker than the bind to other humans. The movement towards the good in more-than-human beings is a movement of beauty and necessity. It is a movement that allows for human ethics and life. It is beautiful in itself and it is beautiful through the human gaze.

This Weilian-inspired ecological ethical system expands the ethical sphere to include more than human life, but not at the expense of human life. The ethical bind to plants, animals, the things of the world is absolutely present and possible, but it does not come at the expense of the social political ethical commitment to other humans. The charge of ignoring the social political in favor of a metaphysical transcendence is not possible in this Weilian-inspired ecological ethics because one cannot choose the preservation of a forest over the lives and livelihood of an Indigenous group who live adjacent to that forest. The things of the world are ethically considered and can call out to us, creating a bind ontologically weaker than our bind to other humans (particularly the most marginalized humans) but still absolutely present and ethically important. Remember that in Weil's reading of Plato's cave, there is *not* an assertion that the things of the world are an illusion, but rather, there is a deepening of our love for those things, a recognition of the Good in our love of the world. Our responsibility towards the world grows in our expansion of the self through decreation, and our recognition of divine love in all things (plants, animals, things). Our ethical duty to and love for the world is bound up in this process. The ethical bind to fellow humans maintains priority, but the ethical bind to the things of the world is absolutely present in this process.

Weil's writing on oppression focuses on industrial work and the force that infuses such work, rendering the worker a mere object. But of course, we now know that poverty-stricken, oppressed humans are not the only casualty of the Industrial Revolution. The surrounding natural world has also been intensely changed, and ecosystems irrevocably annihilated by capitalist industry. By excluding non-human beings from our ethical vision, we have allowed vast portions of our ecological assemblages and root systems to be violently transformed from creative vital life to inanimate object. The ignorance willfully sustained by those who benefit from capitalism and inflicted on those who become its unwitting human accomplices and victims has caused and continues to cause the extinction of innumerable beings (plants, animals, insects, etc.).

In expanding beyond the human-centric limitations of Weil's ethical thought, this is an exercise in thinking with Weil around ecological ethics, but it should not be mistaken for a project found in Weil beyond her use of metaphor. Lissa McCullough notes in her reading of Weil on Judaism that she must read Weil against Weil in order to recognize, through the Weilian pedagogy of decreating and opening space for difference, Weil's own anti-Semitism.[13] Similarly, here I am suggesting to read Weil with, against, and beyond Weil, deepening the decreative process through a Weilian-inspired ecological ethics.

Potential Obstacles to a Weilian Ecology

Opening up Weil's ethical decreative framework to consider the ethical importance of all beings is a logical move from an ecological perspective. When we consider the balance of our earth's ecosystems and the ecological network of which humans are but one small part, we can see how important it is to notice and respect all aspects of this intricate puzzle that is the chaotic, messy earthly realm. Considering human instigators of violence, Weil argues:

> By looking at the world with keener senses than theirs [the great instigators of violence], we shall find a more powerful encouragement in the thought of how these innumerable blind forces are limited, made to balance one against the other, brought to form a united whole by something which we do not understand, but which we call beauty.[14]

Our attention finds necessity as a limit to force that can bring about balance, symmetry, and beauty in all things. In this recognition we find the things and beings of the world participating in order and beauty beyond our full comprehension. To know our ethical duty through our faculties, we must broaden our conception of human community to include ecological assemblage. It is not humans alone that deserve our ethical acts, and humans are not the only beings who show us ethical attention. Indeed, my expansion of Weil's notion of decreation is one that greatly diminishes ontological hierarchy. In particular, I have emphasized Weil's use of *metaxu*, as a way of avoiding a strict hierarchy between real and ideal, which aligns with the type of ecological thinking that Catherine Keller and Mary-Jane Rubenstein note as holding open the distinctions "between spirit and matter, life and nonlife, or sentience and non-sentience."[15] It is the very nature of *metaxu* that both sides of such contraries be held open as one engages in thought and action, mind and body. This lack of hierarchy does not imply a "flat ontology" or a "night in which all cows are black,"[16] or an ontological sameness to all things in which one cannot differentiate anything. The night is perhaps black, but each cow is of a different color and a unique disposition, brimming with diverse bacteria, having eaten different foods, drinking water from a variety of wells and streams, and carrying various flies and insects around her eyes or in his ears. Despite their colors or the quality of light, each cow contains multitudes, and those multitudes are made up of the plants, insects, minerals, humans, viruses, and vitamins that she encounters across the days and years of her life.[17] Keller and Rubenstein note that such an ontology could be "folded, even fractal, but not flat."[18] The ethical expansion I propose does not maintain the absolute hierarchical status of humans, as is the case with Weil's traditionally humanist ethics, but nor does it suggest an absolute dissolution of human dignity in that when all things are equal, human life, particularly the lives of the marginalized, must take priority, as humans alone have the unique dignity of their impersonal desire for the supernatural Good and the divine love that this helps us to experience in the surrounding world, recognizing the beauty and divine dignity of all beings and things. Furthermore, this expansion of moral worth upon the things of the earth is a non-anthropocentric recognition of the being's own potential, with the help of deep ecology's concept of self-realization. I will now turn to how my proposed Weil-inspired ecological ethics can be considered as predominately non-hierarchical and non-anthropocentric in the face of Weil's humanism, religion, and purity.

Justice Beyond *Mere* Humanism

The Weil-inspired ecological ethics that I am setting out affirms the existence of humans as a part of the multitude of the earth. The western humanist subject cannot be separated from the earth or the world in which it moves. It is not absolutely distinct, nor can it be thought of as autonomous from the various life forms that co-create the atomistic human "individual." This subject is human and is situated in the western historical, literary, sociopolitical tradition that makes human life so deeply problematic but also perhaps gives life the possibility of beauty, love, and ultimately meaning. In order to consider what is lost and gained with humanism, it is vital to consider what humanism entails. There are a number of definitions for humanism, and some are specific to historical eras such as ancient Greece, ancient Rome, the Renaissance, or the Enlightenment. Broadly, humanism values an education in classical thinkers, the centrality of human persons, the importance of reason and autonomy, science (rather than religion) as our most powerful tool for knowledge, and moral equality as the foundation for our society. While the claims of total human centrality and autonomy are problematic for this ecological ethics, there is also the positive emphasis on art, culture, literature, philosophy, science, and the celebration of what makes humans worth attending to despite the atrocities of which humans are capable. One need not necessarily believe in the uniqueness or beauty of humanity, and one certainly need not do so through the lens of western humanism. However, there is value in the capacity to marvel at human knowledge, art, and beauty. It provides a way of appreciating the creative possibilities of the human being as a unique entity that has something worth bringing into the fold of an ecological ethics. This is expressed in Weil and in Plato as a divine love or a transcendent recognition of ethical good in the beauty and wonder one experiences in the world. It bears repeating Marie Meaney's explanation that for Weil, as for Plato in his famous ladder of love, beauty "is the only transcendental accessible to the senses, it is the only one which, once inside the cave of Plato's myth, can tempt the prisoners to break their chains and eventually discover the good outside" (*AL*, 63). By transcendental, Meaney is referring to something which "has its origin beyond this world and manifests God" (*AL*, 63). If one does not believe in the human as anything more than a destructive force, as may be suggested in the rejection of humanism full stop, then the risk of nihilism is strong, and the extinction of human beings becomes a preferable outcome. Again, this outcome would begin with the destruction of marginalized communities and expand to more and more people. It would end with the annihilation of all humans, but it would begin with indifference to environmental racism. With this in mind, if one wishes for humanity to continue, there must be something worth redeeming in the human or one's wish is simply an unethical (in regard to the rest of the earth) expression of flagrant self-interest. Just as I propose an ecological reading of Plato over the dominant "anti-materiality" interpretations of Plato, I believe that despite the anthropocentrism and progress-orientation usually noted in humanism by ecological thinkers, the humanist tradition has much to offer ecological ethics by way of affirming the goodness and uniqueness of humanity without ignoring human flaws.[19]

The possibilities of humanism require a necessary ethical stance on what it means to be a human. Farrier notes, "In making kin, we cannot kid ourselves; becoming with others is not a swerve away from the things that mark us out as human but a deepening of these very differences."[20] Humans can contribute to the assemblages of which we are a part through our own process of self-realization and expansion. We can come to recognize how we are similar and also how we are distinct from the other things and beings we encounter. To see this, of course, requires attention. In paying attention to the matter of the world, we see other beings but also ourselves more clearly.

Our focus, if we are aiming towards ethical action, must be towards those excluded and marginalized. For Keller and Rubenstein, ethical thought must engage with ways of thinking beyond reductive totalizing structures such as capitalism, alongside marginalized voices in "feminist, queer, and anti-racist orientations."[21] Myriad counter-capitalist economic possibilities and non-totalizing engagements with others can be considered and enacted as one forms different types of relationships to the surrounding world. Human political and social justice aspirations are not separate from the concern with "natural" or biological aspects of being a part of the larger ecosystem. Environmental racism and gender-based environmental concerns are revealed through ethical attention to the material world. The humanism I wish to affirm differs from classical interpretations insofar as my proposed humanism privileges difference rather than totality or sameness even as it notes what makes the human unique and worthwhile. The human ability to pay attention and to engage with different beings is the highest potential of human beings. Thus, Weil's is a humanism that privileges the science, art, and work oriented towards the human other and mine privileges science, art, and work attentively oriented towards *all* others but with a special place for human dignity.

A Different Kind of Purity

Weil's *metaxu* and her unique mixture of theology and social-justice-oriented materialism is rooted in an ethical appreciation for difference, which saves her from body–mind dualism because the good is always situated in the material and in embodied ethical action. And yet, there remains a troubling concept of purity in her work. Weil speaks of bodily purity as a pathway towards God. This is a spiritual technique utilized by mystics time and again: a simple diet, celibacy, and consumption of only that which is directed towards God. This comes through at times in Weil's writings as a disgust of the body and a desire to purify her own body. Yet, the idea of purity is counterintuitive to the entangled material life that I wish to acknowledge in the Weil-inspired ecological ethics I propose. In this final section of the chapter, I will turn to the inclination towards purity in Weil to explain why ecological ethics cannot maintain certain notions of purity, what Weil's purity entails, and how that can be reconciled with the ecological ethics I am suggesting.

When it comes to ecological ethics, Tsing notes the importance of impurity:

> We are contaminated by our encounters; they change who we are as we make way for others. As contamination changes world-making projects, mutual worlds – and new directions – may emerge. Everyone carries a history of contamination; purity is not an option.[22]

Purity implies that individual things could be extracted or exist apart from the complex relational ecosystem that gives them both life and meaning. Dualistic Platonism assumes that the mind can become pure through the purification of and detachment from the body. Renunciation of material is a method for purification. Broadly speaking, mystical traditions practice cultivating attention through sensual stimulation: consider the use of incense in both the Catholic church and the yogic ashram. In these traditions it is through the sense of smell that a purity (of sorts) is achieved. It is in and through the sensual that the mind focuses attention on the divine. Additionally, mystical traditions use ascetic practices of sensual renunciation, which connect one to God through paring down the distractions of life. These need not be seen as full rejections of the material realm, but rather can be considered as deepening one's sensual pleasure in things that those without attention would barely pause to notice. For example, a bowl of kitchari as one's only meal of the day offers more warmth and pleasure than a rich prime rib if the latter is eaten without lingering to notice the barrage of flavors. Similarly, pausing on a cool winter's day to feel the sun against one's cheek can contain a more erotic caress than a drunken sexual encounter. Renunciation of material things does not necessarily mean a rejection of the sensual, and this is important when considering Weil's notions of purity. For Weil, purity means a pureness of attention as a connection to God. And to connect with God is to connect with matter on a deeper level.

Alexis Shotwell's critique of purity sees the metaphysics of purity as an attempt to separate all things from one another and to harmfully assume that things increase in integrity by being objectively detached from the surrounding world.[23] Purity in this sense is about individuation and not about a Weilian process of decreation. For Weil, purity is the impersonal center of all beings: to purify is to decreate the small egocentric I, allowing for a connection to the core of other beings despite their difference. The purification does not individualize further, but rather it supports a recognition of the impersonal within one's self and shared by all human beings: the desire for the Good. Thus, Weilian purity is not a metaphysics of purity as separation, but a metaphysics of purity as self-abnegation and recognition of the self as interconnected to others. To be pure here is to recognize the impersonal desire for the Good within oneself and thus all things, seeing the purity that longs for the Good at the core of all being. This is an emphasis on the Platonic theory of participation rather than the separate realm of the Forms. Thus, to be pure for Weil is to be entangled and interconnected. Metaphysically, purification for Weil is the exact opposite of the segregation and material rejection of matter that Shotwell critiques. Perhaps a different word may help in conceptualizing the Weilian notion of purity. The purity of the impersonal, decreation and, self-lessness in favor of other beings can be understood as a *fidelity*. This fidelity to radical difference, to the possibility of living ethically, and to the potential within all beings is a fidelity constant through flux. It changes and morphs in each situation; it opens and recedes. It is a lifelong commitment that one must make over again day in and day out to both material bodies and to the interests those bodies may have in their own-most potentials. The fidelity to matter and to God, to gravity and to grace, to virtue and care, is at work in decreation and it is never finished.

An Ecologically Ethical God

Weil's decidedly religious thinking can create a hurdle for ecological thinking of nature-in-itself (and not merely as a means to human ends) if we consider religious thought as entailing a hierarchical division of material and spiritual. Weil's unique interpretation of Plato and the Good gives her a perspective that is distinctly Christian but also distinctly non-dualist, meaning that she avoids the "materiaphobia" that is so often associated with religion according to Keller and Rubenstein.[24] According to Weil's reading of Plato, the Good is only ever accessed in and through engagement with the material. With this in mind, I will now briefly turn to Weil's notion of God and consider how it can work with an ecological ethics.

The moment of total negation of God on earth in Weil's individualized decreative process allows for an ethical critique of one's religious dogmas and practices, as it unsettles these dogmas from absolute power. In Weil, this unsettling is the crashing in of the existential necessities of absurdity, absence, and suffering, which I discuss in the third chapter. The absolute negation of God during the loss of meaning gives rise to a dark night of the soul, the moment of absolute atheism. If one is to hope for mediation towards the Good in this dark absence of God, one must weave theology with ethical, political, and social layers. A belief in God with eschatological certainty cannot be the first step of Weil's decreation. Instead, one begins with the ethical desire to ease the suffering of others. Along the path of ethics, through the recognition of God's absence, from a place of absolute void and without expectation of what God may entail or if God even exists, one experiences God. And while God is the only true end, for Weil, the only true motivation towards God is social justice. As such, all theology must be social, political, and founded upon ethics.

The hope that exists in the void is by no means a hope with eschatological certainty. To imagine Weil as urging blind faith is to misunderstand the decreative process. The darkness is never overcome once and for all. But in this darkness, there is the possibility of the unthinkable God. Weil does not promise a response from God, but a response is not impossible. However, it is almost certain that we will not receive a response from God if we are not open to receive it. Furthermore, "God," as the supernatural spiritual grace that allows us to decreate, is complicated by the reality of material existence and the necessity of force in human lives. This allows Weil to differentiate between the two layers of affliction: one can use one's own affliction through necessity to grow, but one cannot justify the forceful affliction of another person. To confuse affliction with one's response to affliction is never correct, justified, or ethical:

> each phenomenon has two causes, of which one is its cause according to nature, that is, natural law; the second cause is in the providential ordering of the world, and it is never permissible to make use of one as an explanation upon the plane to which the other belongs.
>
> (*IC*, 97)

It is our duty to the other to change the structures that cause her affliction. It is our duty to our self to use the affliction that is inevitable for our own decreation and the expansion of our empathy.

Ecological thinking of the kind I have set out here can embrace Weil's theology as distinct from religious doctrines or certitudes. Indeed, "God" is perhaps one of the simplest and most philosophically robust names for that which is unthinkable. This goes back at least to the Middle Ages, with Anselm's ontological argument that "God is that, than which nothing greater can be conceived."[25] When "God" is conceptualized as a placeholder for wonderous mystery and for our own-most possibilities, It can act as a catalyst towards a probing ethical drive in each person. God, in this sense, is one's self-realization of the Good and the transcendent experience of divine love through recognizing beauty and acting ethically.

The Nietzschean "death of God" marks the death of a certain teleological progress-oriented God rather than the death of God Itself. As such, Friedrich Nietzsche's death of God points to the failure of Platonism in the dualist reading set out in Chapter 7. Platonist dualism has been taken up by Plotinus, Augustine, Aquinas, and baked into Christian doctrine. This Christian Platonism posits the ontological determination of earthly beings from the Good and the subsequent epistemic return to the realm of the Forms through releasing base material existence and embracing the refined spiritual Good. According to Nietzsche, this structure no longer offers adequate meaning to humanity.[26] With the secularization of western thought since the work of thinkers like Nietzsche, Karl Marx, and Sigmund Freud, the notion of "God" has mostly been dropped from the realm of intellectual debate. Similar to the villagers laughing at Nietzsche's mad man for proclaiming the death of God, it is all too easy to miss what has been lost in this death if one focuses on the death of a "supreme being" who represents an epistemology far less logical than atheism. The loss of this figurehead is not what is at issue, but rather, the loss of a meaning structure. Nietzsche's pronouncement of the death of God identifies the death of a particular ontological structure. The loss of this ontological structure is a loss of meaning.

Culturally, the death of God has generally been triumphantly taken up as the death of the mystery (the unknowable origins) and not as the death of progress-oriented teleology and dualism (the ontology). The progress-oriented ontological structure of Christianity has been taken up by liberal Enlightenment humanism in ideals such as colonization, imperialism, and the progress-oriented projects of civilizing and taming the earth and marginalized human communities. (This is obviously quite distinct from the humanism I suggest for an ecological ethics, but it bears noting when we are considering where and how humanism has gone wrong in some cases.) This structure has devolved/developed even further in our current late-stage capitalism. It has dropped the term "God" as the sign of the unsayable that is perhaps the key virtue of religious thought. It has transformed *hope* into a teleological conviction rather than an openness to mystery. The ontological privilege of humans remains in late-stage capitalism as firmly enshrined as ever. Nowhere is it more evident than in the stubborn insistence upon the possibility of the "American Dream" or the neoliberal economic belief that wealth will "trickle down" from the wealthiest to the poorest. These are uncritical hopes that do not ask us to risk anything but rather allow us to move through life with blinders. In many cases, late-stage capitalism has thrown aside the redeeming qualities of religious thinking but has maintained the rotting progress-oriented ontology, dualism, and human exceptionalism.

Following Hent de Vries, Jane Bennett notes that God can be considered to be the "intangible and imponderable recalcitrance of our existence … 'that which tends to loosen its ties to existing contexts'; or the 'limits of intelligibility.'"[27] Perhaps a way to conceptualize God here would be to follow Keller in her assertion of a theology that "keeps faith with its own failures" and "practices a queerer art of theology in the multicolored she/he/it of divinity, the LGBTQ+ they mutability of what we nickname God, and so also in the plurisingular, pancreaturely we of the *imago dei*."[28] Keller encourages one to note but not repeat the failures of one's tradition, practicing mindfulness in the face of history, and staying with the trouble of the unsayable. God, in such a reading is the possibility of standing open before the unsayable. The traditional Christian God will not work with Weil's particular form of negative theological decreation and her social justice-oriented, nondualist materialism, but this traditional God should not be deleted from the scene. Instead, following Donna Haraway's suggestion that the Chthulucene is a process of composting, we must note the death of God and observe what new possibilities can grow from Its corpse.

Luce Irigaray suggests it is not accidental that the hermit whom Nietzsche's Zarathustra finds in the woods has not yet heard of the death of God: "Perhaps, instead of being already dead, the god that we could meet in the forest is still to come."[29] If the God of progress is dead (in that it no longer offers us meaning despite our capitalist insistence that it can and it will), perhaps a new materialist feminist god of the woods is yet to be born. As a theological materialist, Keller suggests that a conception of God need not be thought over and against matter:

> God most certainly fails to meet our expectations, theist, or atheist, of a controlling care. S/he/it might never have meant to succeed so much as to *materialize.* What if the unconditional *matters*? … these are rich old codes of divine mattering in the world, a materialization happening not through control but

> through collaboration. *Creatio cooperativa.* It would be less a matter of a plan being fulfilled than of experiments being enacted.[30]

This God without progress, existing as divine mattering in and through earthly actions, comes closer to the God necessary for a Weilian-inspired ecological ethics. As such, we can see through Weil's non-dualistic reading of Plato and her insistence on an ethics of social justice that there is a place for this God that exists in the love of the earth.

Keller describes Christian negative theology as "a theological unknowing, apophatic theology, in relation to the standard certitudes of theology of Christian exceptionalism."[31] This description certainly holds true for Weil's line of thinking around decreation in which one must unknow the I and reject any certainty. The decreative act of God removing Itself from the earth and in so doing imbuing the earth with the trace of God must be mirrored by individuated people. The spiritual techniques that Weil directs her readers towards are not uncommon mystical tools. Ethically, it deeply matters that God has removed Itself from the earth because this makes it possible for humans to consent to a relationship with God. God cannot morally assert Itself upon the unwilling. Weil explains that "[o]ver the infinity of space and time, the infinitely more infinite love of God comes to possess us. He comes at his own time. We have the power to consent to receive him or to refuse" (*WG*, 78). Speaking of her own political theology of the earth, Keller notes, "To put it with feminist bluntness: We do not embrace a God who forces himself on us. It lures, and the materialization is up to us, in the failings and shifting togetherness of all our entangled becomings."[32] Keller's reading of negative theology raises the possibility of consent and choice. The Weil-inspired ecological ethics I propose invites us to enact the ways in which Weil attempts to unsay, to rub up against the mystery through ethical earthly engagements. This is an attempt to go beyond simply affirming the exciting nature of such mystical claims. It attempts to embody and enact the unsaying through techniques that lead us to ethical action. With this in mind, the final chapter of this book will turn back to the map of our human condition and try to conceptualize how we can think and act ethically in the Anthropocene.

Notes

1 Robin Wall Kimmerer, *Braiding Sweetgrass: Indigenous Wisdom, Scientific Knowledge, and the Teachings of Plants* (Minneapolis: Milkweed Editions, 2013).
2 Anna Lowenhaupt Tsing, *The Mushroom at the End of the World: On the Possibility of Life in Capitalist Ruins* (Princeton: Princeton University Press, 2015), 29.
3 Simone Weil, "Human Personality," in *Simone Weil: An Anthology* (London: Penguin Books, 2005), 66.
4 David Farrier, *Anthropocene Poetics: Deep Time, Sacrifice Zones, and Extinction* (Minneapolis: University of Minnesota Press, 2019), 16.
5 David Abram, *The Spell of the Sensuous: Perception and Language in a More-Than-Human World* (New York: Vintage Books, 2017), 53.

6 Ibid.
7 Lorne Leslie Neil Evernden, *The Natural Alien: Humankind and Environment* (Toronto: University of Toronto Press, 1993), 138.
8 Tsing, *Mushroom*, 141.
9 Arne Naess, "Self-Realization in Mixed Communities of Humans, Bears, Sheep, and Wolves," *Inquiry* 22 (1979), 232.
10 Ibid.
11 See Warwick Fox, "On the Interpretation of Naess' Central Term 'Self-realization,'" *The Trumpeter* 7.2 (1990), 99.
12 See Chapter 3's section "An Ethical Critique of Weil's Force" for more.
13 Lissa McCullough, "The Political Import of Weil's Religious Turn," at *Hannah Arendt and Simone Weil Conference*, April 16–18, 2021.
14 Simone Weil, *The Need for Roots*, trans. Arthur Wills, intro. T.S. Eliot (New York: Routledge 1952), 11.
15 Catherine Keller and Mary-Jane Rubenstein, eds. "Introduction: Tangled Matters" in *Entangled Worlds: Religion, Science, and New Materialisms* (New York: Fordham University Press, 2017), 2.
16 This is Hegel's critique of Schelling's "absolute." See Georg Wilhelm Friedrich Hegel, *Phenomenology of Spirit*, trans. A.V. Miller (Oxford: Oxford University Press, 1977), 9.
17 "Coevolution with microbial life has shaped the 'we' we think we are in unobserved but vital ways for the duration of human history, enfolding different expressions of deep time in our bodies. In noting the depth and range of what she calls 'bacterial liveliness,' Myra Hird recognizes the deep strangeness we find within ourselves when we encounter our bacterial actants: all life forms, she says, 'are, both ancestrally and currently, literally made up of bacteria'; in encountering bacteria, we must recognize that they 'precede my relating with them … that "I" am bacteria, that bacteria are us.'" Farrier, 33. Quoting from Myra J. Hird, *The Origins of Sociable Life: Evolution after Science Studies* (Basingstoke: Palgrave Macmillan, 2009), 22, 51, 26.
18 Keller and Rubenstein, 2.
19 For more on the discords between humanism and ecology from a continental philosophy perspective, see Heather I. Sullivan and Bernhard F. Malkmus, "The Challenge of Ecology to the Humanities: An Introduction," *New German Critique* 43.2 (2016), 1–20. For more on humanism, see John C. Luik, "Humanism," *Routledge Encyclopedia of Philosophy*. https://doi.org/10.4324/9780415249126-N025-1.
20 Farrier, 122.
21 Keller and Rubenstein, "Introduction: Tangled Matters," 3.
22 Tsing, *Mushroom*, 27.
23 Alexis Shotwell, *Against Purity: Living Ethically in Compromised Times* (Minneapolis: University of Minnesota Press, 2016), 16.
24 Ibid., 6.
25 Anselm, *Proslogion*, "Chapter III," in *Readings in Medieval History: Volume II The Later Middle Ages*, ed. Patrick Gleary (Toronto: University of Toronto Press, 2016), 295.
26 Friedrich Wilhelm Nietzsche, "The Madman," in *The Gay Science*, trans. Thomas Common (Mineola: Dover Publications, 2006), 90–91.
27 Jane Bennett, *Vibrant Matter: A Political Ecology of Things* (Durham: Duke University Press, 2010), 3. Citing Hent de Vries, "Introduction," in *Political Theologies: Public Religions in a Post-secular World*, ed. Vries and Lawrence Sullivan (New York: Fordham University Press, 2006), 22.

28 Catherine Keller, *Political Theology of the Earth: Our Planetary Emergency and the Struggle for a New Public* (New York: Columbia University Press, 2018), 119.
29 Luce Irigaray and Michael Marder, *Through Vegetal Being: Two Philosophical Perspectives* (New York: Columbia University Press, 2016), 72.
30 Keller, *Political Theology*, 124.
31 Ibid., 18.
32 Ibid., 147.

10 Action in the Anthropocene

The map of Weil's ethics based on her unique interpretation of Plato and the proposed Weilian-inspired ecological ethics will act as the ground for an application of ethical action in the Anthropocene. While in a sense this is the culmination of the project, it is only suggestive of myriad more possibilities not established here. Weil's rich thought cannot be boiled down into a few acts or exercises, but in light of Weil's dedication to ethical action, I offer some possible ways to begin this decreative process. As Benjamin Davis notes, Weil herself always tested her ideas in the field and reworked them accordingly.[1] Davis argues that Weil "never rested content with asking questions. She also strove to essay her ideas in the world as well as on the page."[2] So, this chapter offers inspiration for mediation and moving from the page to the act. To practice decreation actively is to experience the world phenomenologically with an openness and an awareness of what each particular situation demands, and to respond accordingly. As Simone Kotva notes, the notion of attention has been a popular theme in "the latter half of the twentieth century" for ecological thinkers, but it also has deep roots in the early Christian tradition of "*phusike theoria*, or the 'contemplation of nature,' an idea popular during the first centuries CE."[3] As such, Weil's attention is a unique contribution to a long tradition of mindful environmental ethics.

Perhaps one of the most important things that one can do to move forward the process of decreation in the Anthropocene is to pay attention. Catherine Keller notes that "[l]osses past and losses to come multiply; but so do the bodies who mind them and refuse their inevitability."[4] It is by minding the losses of our earth that we can make changes, refuse the narrative that those losses were inevitable, and move forward with just action and attempts to do better in the future. Anna Tsing describes the art of noticing as a tool for mushroom hunting. Attention is the practice that finds the mushroom after the landscape has been annihilated, as by the atomic bomb at Hiroshima or in unsustainable logging practices. For Tsing, the art of noticing contains the possibility of rebirth despite the grim recognition of the earth's current state of ecological degradation:

> Industrial transformation turned out to be a bubble of promise followed by lost livelihoods and damaged landscapes. And yet: such documents are not enough.

DOI: 10.4324/9781003449621-14

> If we end the story with decay, we abandon all hope – or turn our attention to other sites of promise and ruin, promise and ruin.[5]

It is through a cultivation of the art of attention that we can find hope and support regrowth while simultaneously recognizing the loss of environmental disaster and destruction. In Weilian terms, to find the mushroom in the barren landscape is to find love in the void. Attention to the earth can reveal love in the void, hope in the darkness, and potentiality in the abyss.

Attention is a necessity because we exist in the consistently changing and adapting earth and nothing is ever stagnant in time, not even our participation in the Good or the Forms.[6] Indeed, how we strive towards the Good will be different in every situation. And the Good that we achieve will be dependent upon where we place our attention. Jan Zwicky claims that "when we pay attention, we can tell that the world is awake, that it means, hugely and richly, all the time."[7] In engaging with Weil's work for an ecological ethics, we must attend to the surrounding world and in so doing shift from a narrative of human dominance to one of intraspecies ethical care and reciprocity. This involves a shift in attention, communication, and openness. Luce Irigaray sets out a process of "conversion" as beginning in silence: "Silence is the origin and the medium that allows us to listen to another living being."[8] From silence, we can listen, hear, see, taste, smell, and touch. Irigaray suggests a sensory education of sorts that operates "by accepting not to grasp but to be touched by the sight of a birch, the song of the wind in the woods, the scent of a rose, the taste of a raspberry or of a peach."[9] This is reminiscent of Weil's suggestion that beauty should be looked at and enjoyed but cannot be totalized or incorporated into one's own self. Instead, it must be allowed to simply *be* (*WG*, 105). Attending to the earth through cultivating the attentive process of decreation can happen in any number of ways depending on one's disposition. Here are a few possible options through which to begin such a process:

(1) Walking along the water or through the woods and attempting to keep your focus on the inhalation and exhalation of the breath. If the mind is particularly active, try counting the breaths or silently repeating: inhale, *I am breathing in*; pause; exhale, *I am breathing out*; pause. Repeat. Feel the breathing move in and out with the flow of the water or wind in the trees. Feel the earth and sky breathing with you.

(2) Find an object to observe. This can be anything: a shell, a rock, a stick, a dead bird, an empty beer can. Sit with the object, either where it lies, or take it to a quieter space. Notice everything about it: its size, color, shape, weight, smell. Allow yourself to be with it. Allow it to be.

(3) Step outside and feel the weather on your face. The sunshine, rain, wind, heat or cool. Feel the earth under your feet. Hard, soft, wet, dry. Feel gravity holding you gently in place as the earth stands firmly beneath.

(4) Join a community garden or grow plants in your home or your outdoor space. Get dirt under your fingers. Sing or talk to the plants. Ask the plants questions. Listen for answers.

(5) Before you eat, consider where the food came from. Read the packaging and consider all the beings that have contributed to it arriving on your plate. With each bite, notice the tastes, feel a sense of interconnection to all the links on the chain that brought this food to your plate. Eat gratitude.
(6) Whenever possible, go barefoot and feel your connection to the earth. Similarly, touch trees whenever possible. Listen with your skin.
(7) When you encounter non-human animals, really notice them. Practice withholding your judgments and observe them. Challenge yourself to allow their difference to come forward: this happens by attending to them. Imagine that each one of them has already achieved the attention and openness you are trying to cultivate.
(8) Consider sitting outside or next to the open window and practicing a poetic repetition to focus the mind. The poem that Weil chose for this practice was *Love (III)* by George Herbert:

Love bade me welcome. Yet my soul drew back
Guilty of dust and sin.
But quick-eyed Love, observing me grow slack
From my first entrance in,
Drew nearer to me, sweetly questioning,
If I lacked any thing.

A guest, I answered, worthy to be here:
Love said, You shall be he.
I the unkind, ungrateful? Ah my dear,
I cannot look on thee.
Love took my hand, and smiling did reply,
Who made the eyes but I?

Truth Lord, but I have marred them: let my shame
Go where it doth deserve.
And know you not, says Love, who bore the blame?
My dear, then I will serve.
You must sit down, says Love, and taste my meat:
So I did sit and eat.[10]

My own inclination is towards Paul Reps and Nyogen Senzaki's translations of "Centering," an ancient Sanskrit manuscript which may be the roots of Zen. Here are a few options that I find helpful:

Imagine spirit simultaneously within and around you until the entire universe *spiritualizes* ...
Consider any area of your present form as limitlessly spacious …
Feel your substance, bones, flesh, blood, saturated with *cosmic essence* …
Suppose your passive form to be an empty room with walls of skin – *empty* …

> In summer when you see the entire sky endlessly clear, *enter such clarity* … Wherever your attention alights, at this very point, *experience* …[11]

(9) Allow yourself to pay attention to traditionally "impure" or "unnatural" areas or things such as garbage dumps, industrial parks, or homeless encampments. Allow yourself to open to the destruction or loss of these places. Wait. Open. Here, too, hope or perhaps beauty will arise. Find the matsutake.
(10) When someone speaks to you on the street, stop. Listen. Look. Look in their eyes. Respond. They are real. Don't miss that. Breathe.

These are a few ways that one can begin the process of attending and decreating. The simplicity of these suggestions is deceptive. They are acts that can take a lifetime to embody. I will now turn to specific examples of how we can cultivate the individual faculties of knowing, loving, and, willing in response to absurdity, absence, and suffering through the cultivation of wisdom, loving God, and consenting by actions in science, art, and work which reflect an ethical ecological commitment.[12] An ecological ethics inspired by Weil requires attention and the movement through this process of decreation in response to the affliction of both humans and more-than-humans.

Scientific Wisdom in the Anthropocene

The faculties of willing, knowing, and loving allow for the subject to make meaning in the world and make sense of the surrounding environment, asserting one's own place within that environment. They are the modes through which we come to understand and further form our shared roots. In the epoch of the Anthropocene, cultivating our faculties ethically means recognizing our interconnection to all beings that constitute the earth. Keller notes that the earth is not that matter or space within which we practice politics and theology, but rather "the teeming sphere of our collectivity."[13] The faculty of knowledge here points towards cultivating our understanding of the natural world and our interdependence with it. This knowledge includes the arability of the land, the importance of honeybees in maintaining a healthy ecosystem, understanding how certain chemicals affect certain plants. This is a knowledge that investigates the natural world with an emphasis on our interconnection and shared roots with flora, fauna, fungi, and all of the earth's myriad ways of being. This is manifest through the action of science and in particular what we would refer to as natural, biological, or environmental science. Through the action of natural science, as it is conceptualized in the Weilian context of knowing and absurdity through wisdom, we are able to shed our notion of strict independence and autonomy and instead realize our interconnection as a part of the ecological sphere of the earth. This mediated Weilian science is distinct from the common conception of science in that it is the action of a scientist who has experienced knowledge, absurdity and now acts through wisdom. It sets as its goal the Good, rather than knowing as an accumulation of more power. Knowledge set to work in science decreates the ego-driven sense of the self into the ecology to which it belongs and to which it is beholden. This act deepens one's own interconnection to other beings and also sees oneself as situated in relation to those beings. While I will not give an absolute answer as to what a Weilian-inspired science in the Anthropocene would look like, I will offer some suggestions, namely, a post-scalable, interdisciplinary, diverse epistemic exploration, which considers the scientist's own influence on and interconnection with her work, and promotes responsibility as the key tenet of research. I will unpack these suggestions below.

Recall from the earlier discussion of Weilian action that this is not a science of absolute objective knowing or logical positivism, as this leads to disastrous consequences for the earth and ethical duties. Instead, it is a science that admits to absurdity. It allows for multiple diverse voices, even when those voices are seemingly contradictory. This Weilian-inspired ethical ecology rejects the disaster of blinkered objective truths but maintains the possibility of complex, situated truths in the midst of our entangled reality. For Weil, modern science exchanges thoughtful comprehension of the order of the world for symbolic control producing merely analytic or empirical success and the technological domination of nature. In Weil's opinion, force is at the helm of modern science. Instead, Weil suggests a science that is infused with wisdom, a spiritual Pythagorean science that both understands the world and acts as a potential *metaxu* (*NB*, 514). Indeed, all three of the ethical actions that Weil will suggest are also potential *metaxu* because in acting ethically, one creates a space for the divine Good.

In the face of forceful absurdity, the faculty of knowledge uses wisdom to practice science despite the fact that our knowledge is always imperfect and incomplete. This type of Weilian science always continues to search and explore the best way to articulate a truth. The existential absurdity in a world of growth and decay, flourishing and destruction, can serve to make us all the more grateful for the knowledge we can attain because we see it as rare and beautiful. We do not cling to these truths but see them as ideas *on the way* to an ethical Good that allow us to find the wisdom of beauty and order in the midst of chaos and disorder. A Weilian-inspired ecology for all its insights into the inter-relationality of everything needs to recognize it cannot model itself on a so-called hard science but needs to include all kinds of different ecological understanding so that it is no longer a "discipline" that forces itself onto places administered by experts, but a "field" where universal laws can be seen and explained from numerous perspectives.

Science is often considered to be "objective fact," as I set out in Chapter 2 when considering the logical positivists, but of course this does not capture Weilian science. When it is conceived as a totalizing epistemology, science is closed off from the mystery of other beings.[14] There is not just one single scientific method. There are myriad diverse ways to do science, and by envisioning only one way, we do our wisdom a vast disservice. Tsing suggests that attention to relationships and narratives are crucial aspects of a life-affirming scientific method:

> To learn anything we must revitalize arts of noticing and include ethnography and natural history. But we have a problem of scale. A rush of stories cannot

> be neatly summed up. Its scales do not nest neatly; they draw attention to interrupting geographies and tempos. These interruptions elicit more stories. This is the rush of stories' power as a science.[15]

Tsing notes that because noticing is not scalable or progress-oriented, it has been relegated to the non-scientific realms of thought.[16] For Tsing, scalability limits "correct ways of knowing" to narrow objective theoretical guidelines of a particular set of questions or a particular scientific method. Scalability requires that one only use knowledge that fits within the scale of a certain project. This means that one cannot move between disciplines, frameworks, or methods. It is essentially the old adage, "Don't compare apples to oranges." For Tsing, this does not allow for meaningful diversity. That which does not work within a given system is excluded, and in this, meaningful diversity is lost.[17] That which is scalable often approaches a project from the same angle as past studies and asks similar questions in similar ways. That which is considered to be non-scalable often comes from non-western traditions and diverse knowledge systems. It is excluded from scientific knowing on the grounds of being non-scalable. Timothy Clarke notes that the ability to read across various scales is absolutely crucial for thinking through the Anthropocene.[18] The scales we must bridge in scientific thinking are beyond interdisciplinary methodology, they are individual and generational, local, national, global, and geological. For Tsing, matsutake mushrooms are a poignant example of post-scalability, or the hope that can arise from the disaster of scientific knowledge as the pursuit of absolute facts and relentless progress. Such a hope emerges through an encounter with radical difference.

According to Weil, we can await this difference and do the preparatory work of attention and ethical action through the faculties and necessities, but we cannot force God or grace, which are the ultimate instances of mystery and difference. God and the grace that brings about God's mystery will be encountered as earthen material manifestations and not as an ideal abstraction beyond the natural world. We have no access to the beyond for Weil and as such, we must await grace and God in material existence. Interestingly, Tsing sees a similar quality in matsutake mushrooms: "Humans cannot control matsutake. Waiting to see if mushrooms might emerge is thus an existential problem. The mushrooms remind us of our dependence on more-than-human natural processes: we can't fix anything, even what we have broken, by ourselves. Yet this need not enforce paralysis."[19] If we continue to assume that the subject is fully autonomous and solipsistic, we may see this loss of absolute control as a weakness. But if we consider the strength and support of the assemblages, in relying on others and existing as a part of a larger ecology, we learn that our strength derives from our ability to be vulnerable and open to other aspects of those assemblages.

Tsing notes that just because something is an assemblage that recognizes post-scalability does not mean that it is ethical. To move beyond the progress-oriented model of scientific method is by no means a guarantee of ethical action. A post-scalability assemblage is a framework "beyond scalability," but it has the potential to be even less ethical than its predecessor. A post-scalable science is necessary for ethics, but it is not sufficient. Scalability limits western science to the dominant western scientific method, and thus the marginalized and unique voices are excluded. But while allowing for all voices to contribute may seem like a step towards epistemic justice, it does *not* mean that this will translate into ethical action. Breaking down the barriers of scalability is just one step in this process and if it is not carefully navigated, one risks sliding into a "post-truth" world in which all scientific authority is lost. For example, adding alternative narratives and questions to our knowledge opens up the possibility of giving vaccinations and disinfecting one's hands the same intellectual weight as prayer and scapegoating of vulnerable populations when navigating a pandemic. It could give equal authority to a religious zealot posting videos online as to an epidemiologist.

With this in mind, I would suggest that such a non-scalable system could begin by considering what Sandra Harding termed "strong objectivity" to replace "scientific objectivity."[20] This concept argues that empirical scientific objectivity is a weak version of objectivity and instead, we must magnify our own social political situatedness within a scientific method to strengthen objectivity. This is best undertaken by getting input from those marginalized by scientific communities:

> But a maximally critical study of scientists and their communities can be done only from the perspective of those whose lives have been marginalized by such communities. Thus strong objectivity requires that scientists and their communities be integrated into democracy advancing projects for scientific and epistemological reasons as well as moral and political ones.[21]

This assists in revealing the biases, viewpoints, strengths, and weaknesses of the scientist, her research, and her scientific community. In addition to noting the situatedness of the scientist and her influence and interconnection on the experiment, a Weilian-inspired scientific method must also consider responsibility. Following Lisa Heldke and Stephen Kellert, the scientific method must always ask, "How can this inquiry be made more responsible?"[22] For Heldke and Kellert, "Inquiry is marked by objectivity to the extent that its participants acknowledge, fulfill and expand responsibility to the context of inquiry."[23] To be responsible in this sense is to respond to others and the surrounding world. It means to continually note the interplay and engagement between beings in the process and to move away from the language of "subject" and "object" in the process of scientific inquiry. Most importantly, it means to acknowledge the larger social political implications of the inquiry, and in a Weilian-inspired scientific method it means to also ask, "Who and what could be harmed in this line of inquiry? Who and what is excluded from ethical consideration?" In this sense, it is a science that seeks out the other to whom we are bound in order to protect and assist that other.

An ethical post-scalable science with strong objectivity must continually open to interconnection across what has been considered separate fields of knowledge. Abram notes that narratives and storytelling are key for understanding people as an interconnected part of the ecosystem or the assemblages of a particular landscape: "to an oral culture, experienced events remain rooted in the particular soils, the particular ecologies, the particular places that give rise to them."[24] For Abram, our language makes comprehensible our landscape and only within our landscape is our language truly comprehensible. As such, science must take into account the stories and myths embedded in landscapes if science is to access a fuller picture in its quest for truths.

From Tsing, we can add the necessity of post-colonial practices. A post-colonial science must consider both the miscommunications (aka the voices and narratives that have been excluded from a particular system because they are not scalable to that system) as well as the unified "proper" system (aka the dominant voices and narratives of a particular system).[25] Using matsutake as her guide, Tsing suggests a science of relationality between species and also with the non-living environment.[26] Quoting Dr. Minoru Hamada, Tsing suggests that such a science would embrace mutualism as a form of love.[27] Each of the mediated actions posited by Weil is motivated by the ethical Good, which saturates the active scientist with a divine love in a Pythagorean sense. There is a motivation for ethical good and divine love, which is expressed through carefully discovering and unpacking the intricacies of existence. This practice, while rare, is not without precedent. In the medical sciences, Gabor Maté is one such practitioner, boldly suggesting that a major part of curing his patients is the act of listening to their stories:

> health and illness are not random states in a particular body or body part. They are in fact, an expression of an entire life lived, one that cannot, in turn, be understood in isolation: it is influenced by – or better yet, it arises from – a web of circumstances, relationships, events, and experiences.[28]

Maté points to individual healing and the scientific study of the particular self but in so doing, he directs attention to the self's position in the larger ecology. This asks for us to recognize our interconnection to the surrounding earth in a decreative process that aims to reassert the self as relational. Maté notes that the

> current medical paradigm, owing to an ostensibly scientific bent that in some ways bears more resemblance to an ideology than to empirical knowledge, commits a double fault. It reduces complex events to their biology, and it separates mind from body, concerning itself almost exclusively with one or the other without appreciating their essential unity.[29]

However, Maté is also quick to point out that this is by no means an invalidation of science, medicine, or the good intentions of doctors; rather it "severely constrains the good that medical science could be doing."[30] Similarly, when considering the natural sciences, a broader view by no means invalidates empirical "hard science" but rather asks that it join into a larger inclusive and lively dialogue with alternative ways of knowing focused on the Good, and with a foundation of ethics rather than inserting ethics as an afterthought.

Tsing notes that scientific questions posed by researchers in various countries often align with the social and political history of that country. Science is always guided by our past and our society's ideas of truth; because of this, Tsing proposes a "cosmopolitan science" that acknowledges the non-scalability of moving between various countries and cultural frameworks but still diligently does the work of thinking between these non-scalable entities. Such a science would also incorporate wisdom from various groups by incorporating vernacular knowledge, common sense, and professional knowledge within the framework of scientific knowledge.[31]

Wisdom in the Anthropocene is a balance of rational knowing and existential absurdity. It is an opening of the self to the truths of natural science and a continual epistemic push to question our knowing, rethink our frameworks, and allow for non-scalable, anti-racist intersections across systems of thought. Wisdom in the Anthropocene takes science, broadly and pluralistically understood, to be the best chance we have for truth(s) and honors the lively nature of truths, allowing for an openness to new possibilities and perspectives. It offers moments of *metaxu* when our ethical actions glimpse the beauty and divine love of other beings.

Artistic Love in the Anthropocene

The faculty of loving is the way in which we make the world meaningful out of love, respect, and awe for the other beings with whom we co-create that world. David Farrier notes that art is an incredibly important vehicle for our phenomenological experience of the Anthropocene:

> Poetry can compress vast acreages of meaning into a small compass or perform the kind of bold linkages that it would take reams of academic argument to plot; it can widen the aperture of our gaze or deposit us on the brink of transformation. In short, it can model an Anthropocentric perspective in which our sense of relationship and proximity (and from this, our ethics) is stretched and tested against the Anthropocene's warping effects.[32]

Love and the artistic creation that love inspires can be given short shrift in "serious" logical thought. But for Keller, love is not a sentimentality but a logical discourse on public responsibility, social justice, and sustainable ecology.[33] In the epoch of the Anthropocene, this extends beyond the love we feel for other humans to the love and connection we feel to the land, air, water, plants, and animals. The faculty of loving here becomes the love of the land of our birth, the land of our ancestors, the land that co-created our language. We see this love perhaps most poignantly in immigrants and refugees, those who are displaced from their homes. There is a vast sense of loss in such a displacement, a loss of world. While the loss of one's homeland cannot be replaced, love for one's adopted land can be cultivated. This love extends to the trees that give us shade from the scorching sun in the summer, create bright colors in the fall, offer kindling for our fires in the winter, and bring us the first signs of life with blossoms emerging in the spring. This is a love that mourns the loss of rare plants, the animals killed at the side of the road, the forests bulldozed for pastures, and the pastures cemented for subdivisions. This is a love that sees the links between loving the stranger, the widow, and the orphan,[34] with loving the animals, plants, and things of the earth; and which sees the links of environmental racism and eco-feminism that tell us we cannot neatly separate our ethical failures towards other people from our broken ethical commitments towards non-human beings. The faculty of loving is extremely painful when we open it up to non-human beings in the Anthropocene. In many ways it becomes a faculty of mourning in that so many of the beings we love, be they animal, vegetable, or mineral, are facing mass extinction. But even without mass extinction, to love the rabbit and to love the fox is to recognize the fact that the fox is never going to stop eating rabbits if it has the chance.[35]

In the Weilian structure set out in Chapters 2 through 5 of this book, the type of love that we have shifts in the process of decreation. It begins with love of the particular, which is rattled by absence, or the loss of a particular thing or person, and it is redirected to love of the universal (divine love) through the particular. Love of the self becomes love of God, namely, love of the particular thing becomes love of the universal desire for the Good embodied within that thing. What is worthy of love is the universal desire for the Good at the core of all humans and the divine love and interconnection to all beings, all things. So, it is the universal within the particular that is loved, and this love continues even in the absence of the particular thing that is loved. In the case of absence, love is redirected through an alternative material particular, but it is always the same type of love: a divine unselfish love for the Good or God.

Love is by no means an easy ethical response. To love the universal in the particular means to reveal oneself as vulnerable and perhaps to love a particular thing that is ungrateful or seemingly unworthy of that love. In those moments, one must remember that it is the universal within the particular that one is loving. This is the shift from love of self to love of God. In this way, the individual remains as a part of the whole and cultivates a deep sense of humility. In his work on community, M. Scott Peck notes:

> While rugged individualism predisposes one to arrogance, the "soft" individualism of community leads to humility. Begin to appreciate each other's gifts, and you begin to appreciate your own limitations. Witness others share their brokenness, and you will become able to accept your own inadequacy and imperfection. Be fully aware of human variety, and you will recognize the interdependency of humanity.[36]

Our individual vulnerability allows us to build a connection with one another.

Love is markedly absent from canonical western political theorists such as Karl Marx, John Locke, Thomas Hobbes, and John Rawls.[37] But it is notably present in religious ethics. Bell hooks and Weil share an affinity for a Christian ethic of love and social justice which sets them apart from many modern canonical western political thinkers. Despite a dissertation on love in St. Augustine, Hannah Arendt struggles with the concept of love as a political possibility because romantic and familial love is a sentiment for the private realm only and as such, for Arendt, love is anti-political, drawing the individual out of a public communal mindset and into a mindset of caring for one's beloved over and against the needs of the larger political community. Arendt theorizes love in opposition to the political.[38] Arendt's position explains the core of why many political thinkers eschew the possibility of love coinciding with politics. It is too personal, too emotional, too individual. But of course, for Weil, this love is of the universal not of the individual so there is the possibility for love to form an ethical foundation for the political. Consider Catherine Pickstock's doxological reading of the subject in Plato's *Phaedrus*. Pickstock asserts that Platonic ethical practice entails an orientation towards liturgical praise of the divine more than an ideal of rational contemplation.[39] Pickstock notes this in Socrates' recantation of his first speech against love and the prayer he offers in praise of Eros. This practice of hymn and praise as an ethical act of love directed towards the Good is also an aesthetic act. It is an art of praise or of making manifest one's love of and desire for the Good. As such, artistic practices are spiritual practices in the Weilian sense.

As Weil notes, art recognizes gravity and necessity, and repeats it in an act of love and openness to grace (*NB*, 486). This means that there is a painful recognition of loss, which is carefully traced and repeated with a divine love, therein making meaning and transcendent ethical potential out of that loss. Art can thus become *metaxu*. For Weil, art is the action par excellence to express our ability to love because it recognizes the beauty, awesomeness, and devastation of that which we love. It holds open our love of the particular in the face of the particular's potential absence. It allows us to express a love of the Good, universal, or divine. By creating art that shows our love of other beings and things, we have the chance to give voice to the relationships that make our world meaningful and as such, to share the sometimes-ineffable feelings of love between beings.

In the Anthropocene, art becomes a way to mourn, to remember, to love, to communicate. It is not just a human love expressed to a non-human being, but at its best, it is a listening and opening to the beings that we share our lives with. As Tsing notes, "[i]t is in listening to that cacophony of troubled stories that we might encounter our best hopes for precarious survival."[40] And thus when we are struck with the absence, which is the zenith of existential inevitability in the Anthropocene, we have art to work through the losses and extinctions that have already happened and that we recognize will happen in the future. Art allows for us to mourn and to educate, to recognize love and connection, and to share in the tremendous pain of the absence that extinction brings about. In mourning, art allows us to sustain attention rather than turn away from that which still needs our attention.

Considering the importance of art in the Anthropocene, Robert MacFarlane suggests that when landscapes are not described, they are not given attention.[41] To describe something is to draw out the details of that thing, to force ourselves to attend to it. This notion ties the practice of attention to the written or spoken word and Farrier uses this to suggest the importance of an Anthropocene poetics. But such a poetics, according to both Farrier and MacFarlane, must recognize its own limits. It may be a part of the cultivation of attention and engagement, but it must not overstate its scope and abilities. Weil's link between science, art, and work assists in this type of humility. Poetic worlding through cultivation of attention is always one piece of a larger process. The poet or artist alone does not create the world. They co-create the world alongside science and work through a recognition of other beings both human and non-human.

Art in and of the Anthropocene is myriad. One example is Land Art (sometimes referred to as Earth Art or Earthworks), the predominately American movement originating in the 1960s, which uses the natural landscape to create site-specific structures, and sculptures. This movement features, for example, the 1970 work *Spiral Jetty* by Robert Smithson, a spiral 1,500 foot long and 15 foot wide made of stones, algae, and other organic materials in the Great Salt Lake of Utah. Land Art invites the viewer to pay attention to the natural world. Smithson's choice of location for this particular artwork is not a typical pastoral scene of idealized natural beauty, but rather a saltwater lake surrounded by industrial remnants and cut off from fresh water by the construction of a causeway ten years earlier. Land Art requires the viewer to enter the natural world rather than entering a gallery space, which quite often purposefully suspends or brackets the outside world. Land Art demands attention to our surrounding phenomenal world or *Umwelt*. In this particular piece, Smithson invites his viewer to appreciate the awesomeness and beauty of the post-industrial landscape, not as opposed to the natural world but as part of it. He draws the eye to the color palette of bright red algae, which flourished and spread unchecked after the construction of the causeway on the Great Salt Lake. He offers a beauty and also a stark vast sense of openness, void, and loss in the landscape and conveys that space to the viewer with a great level of care. Smithson participates in the stark landscape, altering it for years and generations to come. But he also mindfully sees it, offering it to the viewer with love and inviting them to open themselves in order to see it with love.

Contemporary artist Mary Mattingly focuses her work on sustainability, climate change, and displacement. Her art is always a form of activism as well as a multimedia experience of performance, sculpture, and photography. There is a keen despair laced with a palpable sense of possibility in her work. The image entitled *Life of Objects* (Figure 10.1) is from the 2013 *House and Universe* series. The photograph, which was also a live performance piece, is a highly personal look at the artist's patterns of consumption and the people and places across the globe that are affected by her consumption. By literally being weighed down under her own "objects," she asks all consumers in late-stage neoliberal capitalist economies to consider the weight of their consumption habits. The performance aspect of her work often brings the art into her surrounding environment. For example, she pulled this boulder of objects across a local bridge, making visible the refuse which we normally dispose of or ignore in landfills, basements, under beds, or in storage units. This bridge was about to be rebuilt in order to allow for larger shipping cargo to be transported on the waterway below. For Mattingly, her own over-consumption was directly affecting not only herself, and the bridge in her neighborhood, but was interconnected to a vast global economic system. With gravity and love, she pays attention to these items and to the interconnection they create between her, the neighborhood, and the world. She invites the viewer to partake in this attention in the void, to open, to care, and even to love.

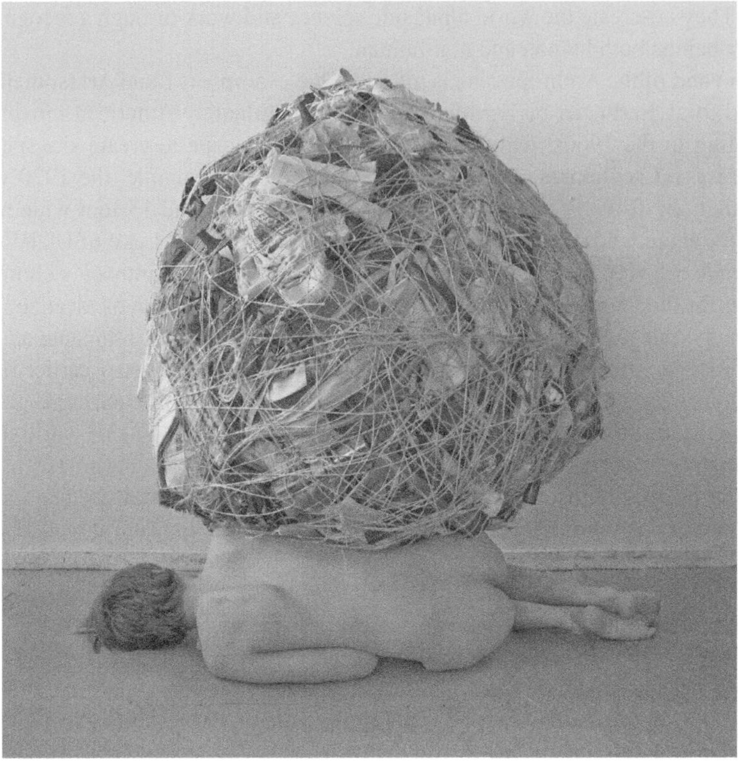

Figure 10.1 Mary Mattingly, *Life of Objects* (2013).

Afrofuturist photographer Fabrice Monteiro's project entitled *The Prophecy* began as a way to raise awareness of environmental degradation and the consequences of excessive consumption in his home of Senegal. The project brings together Indigenous myth, consumerism, and hope for a new future. Haunting, apocalyptic, and yet hopeful images that combine myth and *haute couture*-like design, Monteiro's photographs showcase humans in environmentally devastated landscapes. Monteiro looks directly at the issues of environmental destruction but also celebrates the beauty of the people and the landscape in the face of that destruction. Again, it is a repetition of the gravity of late-stage capitalism that brings love, attention, and beauty to the viewer. One photograph from the series shows a bleak dry landscape in Senegal and a woman perched precariously atop a stack of wood that looks tinder-dry, and the smoke below suggests a fire has already begun. Her metal collar stretching out gives her the look of a windmill, suggesting that this land may have once had water and agricultural possibilities. Now, she is set upon what could be construed as a funeral pyre, looking down, awaiting the fire. The landscape behind her is barren and dry with a gnarled dead tree. But in the distance, there is the hint of green on the horizon. In another photograph from the series, the Australian Great Barrier Reef lies white, bleached, skeletal, dead. A human-like figure, her skin the same bleached white but wrapped in beautiful flowing colors, swims away. The spirit of the Great Barrier Reef is leaving, carrying with it all the life and colors of the reef, leaving behind only the reef's white skeleton.

Dancer and multimedia artist Meagan Woods performed her piece *Slowed Down Sun* on the summer solstice of 2023 in Jersey City overlooking the Manhattan skyline. Woods conceived of the performance, which entailed a single yogic sun salutation slowed down so that it extended across a full hour, as a prayer in motion. From the heart of a major metropolis, Woods' movements invited the bustling world around her to slow, to listen, to recognize the busyness and progress-oriented speed driving the climate crisis. Wearing a white costume of her own design, Woods stitched ice cubes colored by fruit and vegetable juices into pouches that slowly melted across the hour, spreading color and change across her body as it slowly moved through the hot city air (Figure 10.2). An interesting aspect of such slow movement is the grueling training and the strain on both body and mind. Moving quickly is easy and "natural" in our society, but moving slowly and purposefully takes focus and dedication. Woods noted that whenever her mind wandered, or she felt overwhelmed by the process, she would attend to her breath and to the forced movement of climate refugees and would think of her art through the lens of gift-giving culture: she wanted her performance to be a reminder to those displaced by

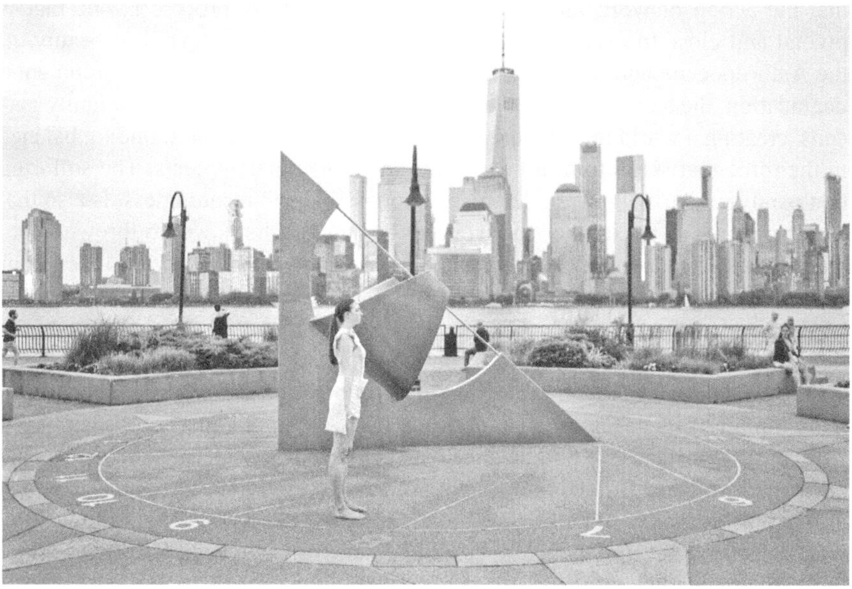

Figure 10.2 Meagan Woods, *Slowed Down Sun*. Photo credit: Lisa Hibbert (2023).

climate-fueled catastrophes that she cared, that others, perhaps across the world, were trying to slow this movement in solidarity. She slowly traced the flow of force with attentive love. Some onlookers joined in, doing sun salutations of their own; others sat and watched in silent contemplation. Still others busily passed along, continuing with their lives as Woods brought prayer into movement through slowing down and attending to the world for that hour.

These are just a small sampling of the potent art of the Anthropocene that practices attention and love while retracing the destruction and inattention of environmental degradation. They are powerful and creative practices of love that invite the viewer into interconnection, responsibility, mourning, love, and action. As I noted, the work of this art has its limitations, but it is a fecund creative expression of gravity and grace that cultivates attention and hope. When paired with science and work, it is ripe with the possibility of ethical action and existence.

Consenting to Work in the Anthropocene

The faculty of willing in the Anthropocene involves a blending of the willpower with attention to the surrounding world. The work of willing is curious within the framework of decreation because the decreative act is precisely about asserting the will in order to release the will. There is a negative attention or openness that happens with this faculty in its decreative mode. Before it is adapted into consenting, the willpower is simply a manifestation of force, or the energy that seeks more and more power. In order to bring balance to force, one shifts from willing to consenting and recognizes necessity and the beauty of the earth. Recall that the action of work that accompanies the decreative process is one that is pivotal and close to divinity, but it can often seem almost devoid of beauty. In the Anthropocene and in regard to issues of climate change and environmental degradation, the action of work may involve grassroots activism, community gardens, creating a world that is more hospitable to other beings, and finding balance in the force that seeks to turn beings into mere inanimate objects. The suffering that paralyzes our willpower, deeming it impossible and meaningless, is also that which inspires us to continue the work. Suffering has the power to throw us into the lower void in an utter despair, but so too does it have the power to incite our action and make us consent to the work before us. *Work* in the Anthropocene must be driven by suffering. If we *love* the earth's beings and *know* our interconnection with said beings, we *consent* to continue, to work against the odds for ethical action. We consent to the imperfection of our world and our work. We consent to reality in all of its necessity and suffering, but we do not consent to accepting it fatalistically. Instead, we find balance between love and suffering and to ease the affliction of others. This work does not hold the beauty of art and science, which in many ways makes it the hardest of the three actions. It is one thing to take on work because one *must* for sustenance and survival; it is another thing to take on work as an ethical duty. This is the type of work that earns Weil her title of "the categorical imperative in skirts."[42] Simone Pétrement gives a description of Weilian work as

> essentially indirect action, action applied to the means. If I stretch my hand toward that which pleases or interests me, this is not work. It is work from the moment that, in order to procure what interests me, I must go toward what does not interest me, make a detour, pass through intermediaries. In real work there is no immediate compensation; to learn how to work is to learn to make an effort in the void, as she would say later on.[43]

Weil's work involved union organizing, socialist and communist agendas, teaching philosophy, geometry, and Greek tragedy to those who were not in a social economic position to receive a higher education, working in factories, fighting fascism with grassroots activism and with the highest levels of government, and attempting to end the oppression of all workers. In the age of the Anthropocene, this work must move beyond simply humans, but it cannot leave humans behind. Instead, it must act alongside art and science to recognize interconnection and love towards all beings. It must pay attention to the needs of the beings most oppressed and diligently attempt to offer relief to those beings.

Contemporary examples of this work are evident in climate activists across the planet. The hubs of this activism are often Indigenous communities and women. I will mention two women who are exemplary in the work they did/do in the Anthropocene: Josephine Mandamin and Naomi Klein. The first example, Josephine Mandamin, was an Anishinaabe First Nations grandmother and elder. I choose to begin with her work because it quite directly affects the land on which I live. Born in 1942 on Wiikwemkoong Unceded Territory, or Manitoulin Island, Mandamin was a founding member of the Water Protectors movement and the Mother Earth Water Walkers, actively working to protect the water in the Great Lakes, on which both she and I live(d). As a survivor of the Canadian residential school system, Mandamin's legacy is a reminder of interconnections of social, political, and environmental issues, which amount to environmental racism and sexism. During her 77 years, she walked around the shorelines of all the Great Lakes, and other North American waterways, carrying a bucket of water, to bring awareness to the need to protect the waters. As a grandmother, a protector of water in the Anishinaabe tradition, Mandamin was the Chief Commissioner of the Anishinabek Nation Women's Water Walk Commission, a member of the Great Lakes Guardian Council. She has inspired generations to come, including Autumn Peltier, her granddaughter, who has taken up a major role as a water warrior since Mandamin's death. Mandamin's work is precisely the type of social, political, environmental work that is necessary for ethical action in the Anthropocene. Her activism draws societal and industrial attention to the issue of clean water, an issue particularly pressing for Indigenous communities under Canadian jurisdiction. According to a Global News report, the Canadian "federal government says 138 long-term drinking water advisories have been lifted since November 2015, although some short-term boil water advisories have also slipped into the long-term category during that timeframe. As of Feb. 3, 2023, there were 32 long-term boil water advisories in 28 communities in Canada."[44] The problem here is not simply the blatant environmental resource racism of this epidemic, but the apathetic and sluggish response time from the federal government. An earlier report from the nonprofit environmental group, the Council of Canadians states that "In 2015, there were drinking water advisories in 126 First Nations. The Trudeau government committed to resolving them by March 2021, but they failed to provide all the funding required to meet that deadline. There are still advisories in 33 First Nation communities."[45] Work in the Anthropocene is necessarily environmental, but our attention reveals that it must also be social, political, and decolonial.

Naomi Klein is a well-known Canadian environmental activist, writer, and academic. University of British Columbia Professor of Climate Justice and co-director of the Centre for Climate Justice, Klein is an example of someone whose work in the Anthropocene happens both within and beyond the academy. Klein's books have drawn attention to the realities of neoliberal capitalism, consumer society, and environmental disaster within academia and in the larger popular culture. The attention that Klein gives to issues of globalization and late-stage capitalism are singular, clear, and incredibly important. They offer a counternarrative to neoliberal capitalist economics.[46] This counternarrative precisely looks at the world rather than simply at theoretical realities. It engages with difference and the reality of affliction in industrial (*No Logo*), social (*Doppelganger*), and political (*The Shock Doctrine*) structures. Furthermore, it pays attention to the radically particular as interconnected and reflective of the universal. In Klein's book on climate change, *This Changes Everything: Capitalism vs. the Climate*, she parallels her own fertility struggles and the planetary age of extinction. She describes her body wading in a British Petroleum oil spill for research, unaware that she is pregnant. She sees the salmon unable to spawn in the polluted water and her own inability to conceive as bound together. She allows her work to be interconnected and to find the universal at the heart of the particular, consenting to necessity and bringing it into balance.[47]

There are myriad examples of work in the Anthropocene, and these are but two women who have been ethically working on Turtle Island in the epoch referred to as the Anthropocene, paying attention to the earth, feeling interconnected with the beings surrounding them, and attempting to bring balance to a tumultuous situation. One step towards this type of work in our own lives is to connect to one of the mutual aid organizations found in many communities around the world. These groups bypass traditional state-sponsored or charity-focused agendas, and act as avenues for mutual respect, community, and support. There is nothing perfect about these organizations, and the work is never complete. It is endless, and as each of these activists would undoubtedly attest, it is often heart-wrenching. This is why such work must be tempered with the beauty and symmetry of art and science. We must continually be reminded why the work is important and the beauty of the void in which we live together. The three actions of science, art, and work allow for the whole of our being to take part in an ethical life. By taking part in all three, we can find balance in ourselves as a part of the larger assemblages of the earth and God. The three actions offer the possibility of beauty, mourning, creation, loss, connection, suffering, joy, and most of all, ethical action in a meaningful life. If one action is emphasized to the exclusion of the other two, we lose balance. Each action is distinct but overlapping.

One question that may arise from these examples of science, art, and work is what the benefit added of the Weilian lens may be, particularly if there are people in the world already doing these things. People like Klein and Mandamin in their work, Woods and Mattingly in their art, or Harding, Heldke, and Kellert in their science are already practicing these mediated modes of ethical ecological being. I suppose this depends on what we take to be the purpose of philosophy. Does it set out new paths for the future? Does it notice the patterns of the past and carefully name them? Is it a gesture to the present and an invitation to open to its possibilities? In a sense, it can be all of these things and I hope that all are embedded in this book. But I do not want to lay any claims to a radical new way of being in the world. Rather, I suggest a patience and openness with ways that have been here for a long time, perhaps since Plato, probably before, certainly in other philosophical traditions such as many Indigenous philosophies and ways of knowing. I think that change is slower than we would like and that our minds and actions change over days, months, and years rather than merely in "Eureka!" moments. As such, this mapping of Weil's philosophy into the sphere of ecological ethics is meant to offer us touchstones as to how we can be ethical in this epoch: how we can grow and change and direct our actions towards environmental health and well-being; how we can stay on an ethical path and allow ourselves to continually grow into new challenges and expand the sphere of ethical importance.

Let us return again to artist and scholar Meagan Woods and consider another one of her projects, this one a collaboration with a number of artists to create an experimental opera/installation entitled *Once She Dries* (Figures 10.3 and 10.4) which tells the story of coral reef and climate change.[48] The multifaceted piece draws together science, art, and work in a way that we may consider through this Weilian lens. First and foremost, this project is a love story based in natural science. It is the love between coral, warming, heating, drying, and the cloud that she forms above from her chemical dimethyl sulfide. Their love is disrupted by human creation and industry, namely the creation of the marvelous structure of the Pantheon in Rome, a building that houses the Gods and represents the human use of the natural world as a resource, or a love that abducts, that does not ask for permission, that demands to be loved in return. The foundation of this story is natural science in that it notices and seeks to understand the relationship between coral and cloud and the disruption in this relationship brought on by anthropogenic climate change. It begins in science. This tale uses art to draw together mythos and logos:

> The libretto of *Once She Dries* tells the story of a coral reef in peril, in desperate need of shade and cooling. The reef's plaintive cry ascends from the ocean floor and releases a cloud-producing vapor. The cloud is charged with finding the nourishment the reef requires. Their bond – the coral, the cloud – and their commitment to each other will bring healing. It is an expression of love. But the way forward isn't without treachery. All the gods have been eavesdropping. The largest cast concrete dome, home to all the gods – the Pantheon – has become the keeper of the reef. (Coral was once mined for lime to give cement

its solidity.) Coral has been trapped under the dome, like Danaë shut up in her bronze chamber, or Eurydice consigned to Hades.[49]

It is immediately clear that this piece draws together science and art in the mediated Weilian sense. Science is not separated from myth, but rather, included into our mythology in the same way Jennifer Rapp, Farrier, Tsing, and others have suggested that mythos forms the world that we walk into and the ways we question that world scientifically.

The artistic team behind the project have "teased out a tale that recognizes love as its ideal while embracing science as a divine entity. The cosmos swirls in a coral reef and we will be judged on how we proceed."[50] In addition to the powerful libretto by Woods, there is the haunting opera itself by pianist and composer Casper Leerink and violinist and composer Kourosh Ghamsari-Esfahani; multi-video projections that tell the story through images of coral, sea, dance, sculpture, and ethereal moments in between that move from fever dream to warm memory by filmmaker, photographer, and installation artist Xinyue Liu; vocal musical contributions from musician and actor Amanda Sum; and the entire space is adorned with delicate paper sculptures of coral, sea, and sky by sculptor and installation artist Nancy Cohen. The entire work has a strange and beguiling sense that combines the radical otherness of the sea with the familiarity of love lost and human heartbreak. Each element of this work enacts the double descent that Weil requires of art in that it lovingly retraces

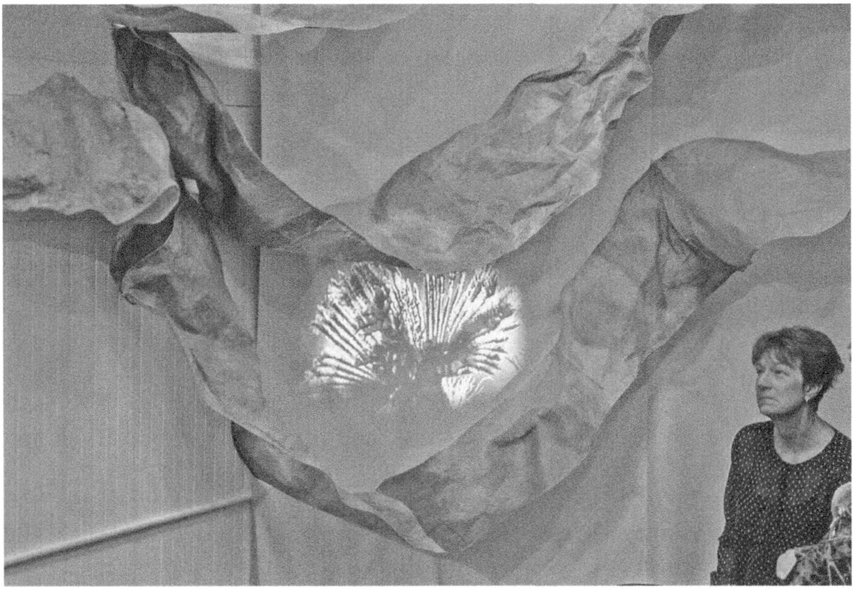

Figure 10.3 Meagan Woods et al., *Once She Dries*. Segment of handmade paper with video projection (2022). Photo credit: Jennifer Brown (2023).

the work of gravity and force through grace and love. It invites attention to coral (the sea), to cloud (the weather systems), to Pantheon (human construction and gods), and as in all love stories, it asks the viewer to consider that which is in between these elements, that which binds and rends them.

The final element of Weilian work manifests in and around the individual presentations and installations of the piece. Woods notes that it is of the utmost import to engage the specific community welcoming the work and to interweave the art with the community rather than enforcing its ideals upon that community, and with that in mind, each exhibition will be unique depending on the needs of the community. Local and international environmental organizations are promoted at the installation, activism is encouraged around the art. Woods screened a documentary on coral, brought in talks around the intersections of science and art, put on potlucks and youth workshops, and perhaps most importantly and time-consumingly, she was simply present for others as they walked through the exhibit. Woods notes that taking the time to slow down, to listen to those who have attended the installation and surrounding events, has been one of the most important aspects of the work. She differentiates this work from her art, in that the art simply *is* in itself. It need not and should not be activism, in her opinion, but of course the implications of this love story make room for the possibility of work. The activism, which engages with the science that inspired the art, is related to but unique from the art in itself. All three actions intertwine but are distinct. In all of its instantiations,

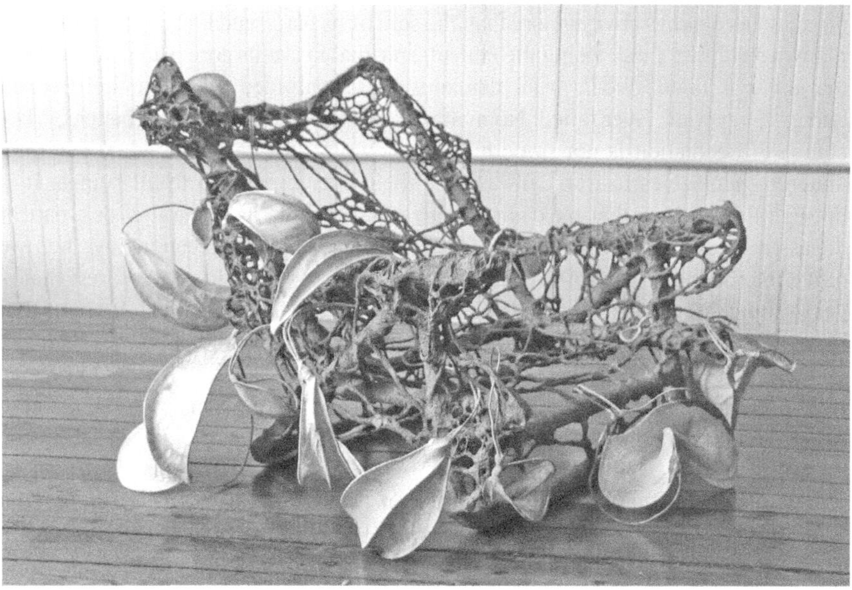

Figure 10.4 Nancy Cohen et al., *Once She Dries*. Coral chair with handmade paper, wire, thread, etc. (2022). Photo credit: Maddie Orton (2023).

the piece seeks to allow people to both know and feel the realities of climate change together: to see our impact and our interconnection to the surrounding world.

Reflecting on her work on vibrant matter, Jane Bennett notes that the political project she is pursuing is "to encourage more intelligent and sustainable engagements with vibrant matter and lively things."[51] Indeed, the epistemic act of reimagining our ontological relationship to matter is absolutely a political act. The actions of decreation are never focused narrowly on oneself but on one's relationships in the complex ecosystem of which one is a vital part. The age of the Anthropocene is one that is rife with absurdity, suffering, and absence. Thus, Weil's practice of decreation through the cultivation of knowing, loving, and willing while holding these faculties in balance with existential force is an incredibly pertinent structure through which we can think, feel, and ethically act in the Anthropocene.

Learning to Die as Actively Living

The becoming of decreation, the process of learning to die as an individuated self in order that one may live in relation to other beings, both familiar and radically other, is a dauntingly personal project. But this is the demand of Simone Weil. We cannot simply think. We must love and act thoughtfully. As the Anthropocene crashes onward through the twenty-first century, it becomes painfully clear that it is only gaining momentum, and that the formidable powers of capitalism and progress fuel its trajectory.

The danger of mapping out Weil's work with these specific examples is that it risks losing the deeply nuanced transcendent and mysterious aspects of Weil's thought: her work is always more than the sum of its parts, and I fear that in breaking it down as I have, I risk losing the ethical, spiritual fire at its core. But again, I think that this risk is one Weil's work demands. I am reminded of the Samuel Beckett quote: "Ever tried. Ever failed. No matter. Try Again. Fail again. Fail better."[52] The course of decreation is not smooth; it is a journey that each individual can only undergo themselves to ironically realize their interconnection to all beings. It is through this journey that we can recognize our position as a plurality or a part of the assemblages of God and nature. From this place of self-decreation, together, we move into decreating the Anthropocene and awaiting the God of the forest that is yet to come. We await and invite this God through our ethical actions. It seems that in reality, the whole is always out of our hands, yet we are the only ones who can manifest the Good in our particular human way. We cannot believe in God if one child suffers. So, we must ease the suffering of that child, of that animal, of these beings and things upon the earth if we wish to flourish and participate in the Good.

> The beauty of the world is the mouth of a labyrinth. The unwary individual who on entering takes a few steps is soon unable to find the opening. Worn out, with nothing to eat or drink, in the dark, separated from his dear ones, and from everything he loves and is accustomed to, he walks on without knowing anything or hoping anything, incapable even of discovering whether he is really going forward or merely turning round on the same spot. But this affliction is

> as nothing compared with the danger threatening him. For if he does not lose courage, if he goes on walking, it is absolutely certain that he will finally arrive at the center of the labyrinth. And there God is waiting to eat him. Later he will go out again, but he will be changed, he will have become different, after being eaten and digested by God. Afterward he will stay near the entrance so that he can gently push all those who come near into the opening.
>
> (*WG*, 103)

In this time of the Anthropocene, when we need it more than ever, may we allow Weil's firm words to push us into the labyrinth of ethical ecological existence.

Notes

1 Benjamin P. Davis, *Simone Weil's Political Philosophy: Field Notes from the Margins* (Maryland, Rowman and Littlefield, 2023).
2 Ibid., 15.
3 Simone Kotva, *Effort and Grace: On the Spiritual Exercise of Philosophy* (London: Bloomsbury, 2020), 174–175.
4 Catherine Keller, *Political Theology of the Earth: Our Planetary Emergency and the Struggle for a New Public* (New York: Columbia University Press, 2018), 158.
5 Anna Lowenhaupt Tsing, *The Mushroom at the End of the World: On the Possibility of Life in Capitalist Ruins* (Princeton: Princeton University Press, 2015), 18.
6 See Chapter 7 on *metaxu* and Plato for more of the realm of the Forms and Plato's theory of participation.
7 Jan Zwicky, "Lyric Realism: Nature, Poetry, Silence, and Ontology," Green Imagination issue, *The Malahat Review* 165 (Winter 2008).
8 Luce Irigaray and Michael Marder, *Through Vegetal Being: Two Philosophical Perspectives* (New York: Columbia University Press, 2016), 49.
9 Ibid., 51.
10 George Herbert, *George Herbert and the Seventeenth-Century Religious Poets*, ed. Mario Di Cesare (New York: W.W. Norton, 1978). See my earlier discussion of this in Chapters 1 and 4.
11 Paul Reps and Nyogen Senzaki, eds., *Zen Flesh, Zen Bones: A Collection of Zen and Pre-Zen Writings* (Rutland: Tuttle Publishing, 1985), 210, 214, 220.
12 Please refer to Figure 2.1 for a visual aid.
13 Keller, *Political Theology*, 6.
14 This critique of the sciences as "objective truth" has been engaged with by many philosophers. For more on the this, see Alan Francis Chalmers, *What Is This Thing Called Science?* (St. Lucia: University of Queensland Press, 1976); Steven Shapin, *Never Pure: Historical Studies of Science as If It Was Produced by People with Bodies, Situated in Time, Space, Culture, and Society, and Struggling for Credibility and Authority* (Baltimore: Johns Hopkins University Press, 2010); Steven Shapin, *A Social History of Truth: Civility and Science in Seventeenth-Century England* (Chicago: University of Chicago Press, 1994); Michael Zimmerman, *Heidegger's Confrontation with Modernity: Technology, Politics, and Art* (Bloomington: Indiana University Press, 1990). Notably, this critique is also engaged with by feminist philosophers.
15 Tsing, *Mushroom*, 37.
16 Ibid., 38.

17 For more on this, see Anna Tsing, "On Nonscalability: The Living World Is Not Amenable to Precision-Nested Scales," *Common Knowledge* 18.3 (2012), 505–524.

18 Timothy Clarke, "Derangements of Scale," in *Telemorphosis: Theory in an Era of Climate Change*, ed. Tom Cohen (London: Open Humanities Press, 2012), 152. Referenced in David Farrier, *Anthropocene Poetics: Deep Time, Sacrifice Zones, and Extinction* (Minneapolis: University of Minnesota Press, 2019), 20.

19 Tsing, *Mushroom*, 257.

20 See Sandra Harding, "Rethinking Standpoint Epistemology: What Is 'Strong Objectivity?'" in *Feminist Theory: A Philosophical Anthology*, ed. Ann E. Cudd and Robin O. Andreasen (Oxford: Blackwell Publishing, 2005). For more, see Sandra Harding, *Whose Science? Whose Knowledge? Thinking From Women's Lives* (Ithaca: Cornell University Press, 1991).

21 Harding, "Rethinking Standpoint Epistemology," 459.

22 Lisa M. Heldke and Stephen H. Kellert, "Objectivity as Responsibility," *Metaphilosophy* 26.4 (October 1995), 360–378.

23 Ibid., 361.

24 David Abram, *The Spell of the Sensuous: Perception and Language in a More-Than-Human World* (New York: Vintage Books, 2017), 162.

25 Tsing, *Mushroom*, 218.

26 Ibid. See also Evelyn Fox Keller, *A Feeling for the Organism: The Life and Work of Barbara McClintock* (San Francisco: W.H. Freeman, 1983).

27 Tsing, *Mushroom*, 220.

28 Gabor Maté and Daniel Maté, *The Myth of Normal: Trauma, Illness, and Healing in a Toxic Culture* (Toronto: Penguin Random House, 2022), 9.

29 Ibid., 8.

30 Ibid.

31 Tsing, *Mushroom*, 263.

32 Farrier, *Anthropocene Poetics*, 5.

33 Keller, *Political Theology*, 157.

34 The widow, orphan, and child are those for whom Levinas argues we are ethically responsible. Emmanuel Levinas, *Totality and Infinity: An Essay on Exteriority*, trans. Alphonso Lingis (Pittsburgh: Duquesne, 2013), 77.

35 For more on ecological mourning, see Ashlee Cunsolo and Karen Landman, eds. *Mourning Nature: Hope at the Heart of Ecological Loss and Grief* (Montreal: McGill-Queen's University Press, 2017).

36 M. Scott Peck, *The Different Drum: Community Making and Peace* (New York: Touchstone, 1998). Quoted in bell hooks, *Belonging: A Culture of Place* (Abingdon: Routledge, 2008), 67.

37 Michael J. Monahan, "Emancipatory Affect: bell hooks on Love and Liberation," *The CLR James Journal: Special Issue On the Emancipatory Thought of bell hooks* 17.1 (2011), 102–111.

38 Hannah Arendt, *The Human Condition* (Chicago: University of Chicago Press, 1998), 51.

39 Catherine Pickstock, *After Writing: On the Liturgical Consummation of Philosophy* (Oxford: Blackwell, 1998), 4.

40 Tsing, *Mushroom*, 34.

41 Robert MacFarlane, *Landmarks* (London: Hamish Hamilton, 2015), 27. Referenced in Farrier, 4.

42 Jacques Cabaud, *Simone Weil: Fellowship in Love* (London: Harvill Press, 1964), 36.

43 Simone Pétrement, *Simone Weil: A Life*, trans. Raymond Rosenthal (New York: Pantheon Books, 1976), 61–62.
44 Marney Blunt, "'We Have to Fix It Faster': 28 First Nations Communities Still Under Boil Water Advisories" *Global News*, March 22, 2023 (online: https://globalnews.ca/news/9571066/first-nations-drinking-water-issues-world-water-day-2023/).
45 "Safe Water for First Nations," *Council of Canadians* (online: https://canadians.org/fn-water).
46 This style of economics is championed by Milton Friedman and the Chicago School of Economics. This school of thought has been highly influential, but it is also theoretically and empirically critiqued. See, for example, Edward Nell, ed., *Free Market Conservatism: A Critique of Theory and Practice*, 1st ed. (Florence: Taylor and Francis Group, 2009).
47 Naomi Klein, *This Changes Everything: Capitalism vs. the Climate* (New York: Knopf Canada, 2014), 434–435.
48 The group includes Nancy Cohen, Kourosh Ghamsari-Esfahani, Casper Leerink, Xinyue Liu, Amanda Sum, and Meagan Woods.
49 Raymond E. Mingst and Arthur Bruso, "Science, Myth, Life, and Love," in *Digital Once She Dries Catalogue* (online: www.onceshedries.com/digital-catalog).
50 Ibid.
51 Jane Bennett, *Vibrant Matter: A Political Ecology of Things* (Durham: Duke University Press, 2010), viii.
52 Samuel Beckett, *Worstward Ho!* (New York: Grove Press, 1983).

Bibliography

Abram, David. *The Spell of the Sensuous: Perception and Language in a More-Than-Human World*. New York: Vintage Books, 2017.

Agamben, Giorgio. *Homo Sacer: Sovereign Power and Bare Life*. Translated by Daniel Heller Roazen. Stanford: Stanford University Press, 1995.

Al-Khalili, Jim. *The House of Wisdom: How Arabic Science Saved Ancient Knowledge and Gave Us the Renaissance*. London: Penguin, 2012.

Allen, Diogenes. "George Herbert and Simone Weil." *Religion and Literature* 17, no. 2, *Simone Weil* (1985): 17–34.

Andic, Martin. "Simone Weil and Kierkegaard." *Modern Theology* 2, no. 1 (1985): 20–41.

———. "Supernatural Justice and the Madness of Love." *Cahiers Simone Weil* 17 (1994): 373–405.

Andrew, Edward. "Simone Weil on the Injustice of Rights-Based Doctrines." *Review of Politics* 48, no. 1 (Winter, 1986): 60–91.

Angelou, Maya. "Alone." In *Oh Pray My Wings Are Gonna Fit Me Well*. New York: Random House, 1975.

Anselm. *Anselm's Basic Writings*, 2nd edition. Translated by S.W. Deane. La Salle: Open Court, 1962.

Aquinas, Thomas. *The Hackett Aquinas: Basic Works*. Edited by Jeffrey Hause and Robert Pasnau. Indianapolis: Hackett, 2014.

Arendt, Hannah. *The Human Condition*. Chicago: University of Chicago Press, 1998.

———. "Thinking and Moral Considerations: A Lecture." *Social Research* 51, no. 1/2 (1984): 7–37.

Aristotle. *Nicomachean Ethics*. Translated by Terence Irwin. Cambridge: Hackett, 1999.

Asdal, Kristin. "The Problematic Nature of Nature: The Post-Constructivist Challenge to Environmental History." *History and Theory: Special Issue on Environment and History* 42, no. 4 (2003): 60–74.

Augustine. *Confessions*. Translated by Henry Chadwick. Oxford: Oxford World Classics, 2008.

Aurelius, Marcus. *The Meditations*. Translated by George Long. MIT, 1994. Online: http://classics.mit.edu/Antoninus/meditations.html.

Barad, Karen. *Meeting the Universe Halfway: Quantum Physics and the Entanglement of Matter and Meaning*. Durham: Duke University Press, 2007.

Barker, Ernest. *The Political Thought of Plato and Aristotle*. New York: Franklin Classics, 1959.

Bartlett, Christopher. "Nautilus Spirals and the Meta-Golden Ratio Chi." *Nexus Network Journal* 21 (2019): 641–656. https://doi.org/10.1007/s00004-018-0419-3.

Bataille, Georges. *The Accursed Share: Vol I.* Translated by Robert Hurley. Brooklyn: Zone Books, 1989.

Beckett, Samuel. *Waiting for Godot.* New York: Grove Press, 1982.

———. *Worstward Ho!* New York: Grove Press, 1983.

Bennett, Jane. *Vibrant Matter: A Political Ecology of Things.* Durham: Duke University Press, 2010.

Bonneuil, Cristophe and Jean-Baptiste Fressoz. *The Shock of the Anthropocene.* London: Verso, 2015.

Brewer, Holly. "Slavery, Sovereignty, and 'Inheritable Blood': Reconsidering John Locke and the Origins of American Slavery." *American Historical Review* 122, no. 4 (2017): 1038–1078.

Brindle, Wayne A. et al., editors. *The Holy Bible. New King James Version.* Nashville: Thomas Nelson, 1988.

Brundige, Lorraine and Douglas Rabb. "Phonicating Mother Earth: A Critique of David Abram's *The Spell of the Sensuous: Perception and Language in a More-Than-Human World.*" *Ayaangwaamizin: The International Journal of Indigenous Philosophy* 1, no. 2 (1997): 78–88.

Burger, Ronna. "Socratic Irony and the Platonic Art of Writing: The Self-Condemnation of the Written Word in Plato's 'Phaedrus.'" *Southwestern Journal of Philosophy* 9, no. 3 (1978): 113–126.

Burns, Robert. *The Complete Poems and Songs of Robert Burns.* Glasgow: Gresham, 2015.

Burns, Steven. "Justice and Impersonality: Simone Weil on Rights and Obligations." *Laval théologique et philosophique* 49, no. 3 (1993): 477–486.

———. "Virtue and Necessity." *Laval théologique et philosophique* 32, no. 3 (1976): 261–275.

Butler, Judith. *The Force of Nonviolence.* New York: Verso, 2020.

Cabaud, Jacques. *Simone Weil: Fellowship in Love.* London: Harvill Press, 1964.

Cajori, Florian. *A History of Mathematical Notations.* New York: Dover, 2011.

Callicott, J. Baird. "Intrinsic Value, Quantum Theory, and Environmental Ethics." *Environmental Ethics* 7 (1985): 299–309.

Cameron, Sharon. "The Practice of Attention: Simone Weil's Performance of Impersonality." *Critical Inquiry* 29, no. 2 (2003): 216–252.

Camus, Albert. *The Myth of Sisyphus.* London: Penguin Books, 2005.

Cantini, Andrea and Riccardo Bruni. "Paradoxes and Contemporary Logic." *The Stanford Encyclopedia of Philosophy* (Fall 2021). Edited by Edward N. Zalta. Online: https://plato.stanford.edu/archives/fall2021/entries/paradoxes-contemporary-logic/.

Caranfa, Angelo. "The Aesthetic and the Spiritual Attitude in Learning: Lessons from Simone Weil." *Journal of Aesthetic Education* 44, no. 2 (2010): 63–82.

Carson, Anne. "Shoes: An Essay on How Plato's 'Symposium' Begins." *Iowa Review* 25, no. 2 (1995): 47–51. www.jstor.org/stable/20153663.

Cha, Yoon Sook. *Decreation and the Ethical Bind: Simone Weil and the Claim of the Other.* New York: Fordham University Press, 2019.

Chalmers, Alan Francis. *What Is This Thing Called Science?* St. Lucia: University of Queensland Press, 1976.

Clarke, Adele E. and Donna Haraway, editors. *Making Kin Not Population.* Chicago: University of Chicago Press, 2018.

Clarke, Timothy. "Derangements of Scale." In *Telemorphosis: Theory in an Era of Climate Change*, 148–166. Edited by Tom Cohen. London: Open Humanities Press, 2012.
Cockburn, David. "Simone Weil on Death." *Mortality* 2, no. 1 (1997): 63–72.
Colebrook, Clare. *Anthropocene Feminism*. Edited by Richard Grusin. Minneapolis: University of Minnesota Press, 2017.
Corbí, Joseph E. "First Person Authority and Self-Knowledge as an Achievement." *European Journal of Philosophy* 18, no. 3 (2009): 325–362.
Cosgrove, Joseph K. "Simone Weil's Spiritual Critique of Modern Science: An Historical-Critical Assessment." *Zygon* 43, no. 2 (2008): 353–370.
Council of Canadians. "Safe Water for First Nations." Online: https://canadians.org/fn-water.
Cronon, William. "The Trouble with Wilderness; or, Getting Back to the Wrong Nature." In *Uncommon Ground: Rethinking the Human Place in Nature*, 69–90. Edited by William Cronon. New York: W.W. Norton, 1995.
Crutzen, Paul J. "Geology of Mankind." *Nature* 415 (2002). https://doi.org/10.1038/415023a.
Culler, Jonathan. *Theory of the Lyric*. Cambridge, MA: Harvard University Press, 2015.
Cunsolo, Ashlee and Karen Landman, editors. *Mourning Nature: Hope at the Heart of Ecological Loss and Grief.* Montreal: McGill-Queen's University Press, 2017.
Dalai Lama XIV. *The Four Noble Truths: Fundamentals of Buddhist Teachings*. Translated by Geshe Thupten Jinpa. Edited by Dominique Side. London: Thorsons, 1997.
Davis, Benjamin P. *Simone Weil's Political Philosophy: Field Notes from the Margins*. Lanham: Rowman and Littlefield, 2023.
Davison, Andrew. "The Mediating Possibilities of Absence in the Thought of Simone Weil." *Theology* CXII.865 (2009): 3–13.
Derrida, Jacques. "Différance." In *Margins of Philosophy*, 3–27. Translated by Alan Bass. Chicago: University of Chicago Press, 1982.
———. "Plato's Pharmacy." In *Dissemination*, 63–171. Translated by Barbara Johnson. Chicago: University of Chicago Press, 1981.
———. "Restitutions of the Truth in Pointing." In *The Truth in Painting*, 255–382. Translated by G. Bennington and I. McLeod. Chicago: University of Chicago Press, 1978.
Descartes, René. *Meditations on First Philosophy*. Edited by Andrew Bailey. Translated by Ian Johnston. Toronto: Broadview Press, 2013.
Desmond, William and Kevin Kennedy. "The Theater of the *Metaxu*: Staging the Between." *Topoi* 30, no. 2 (2011): 113–124.
Devall, Bill. "Issues in Contemporary Ecophilosophy." Paper presented to the "Ecology and Society Conference," University of Wisconsin at Waukesha, April 7, 1984.
———. *Simple in Means, Rich in Ends: Practicing Deep Ecology*. Layton: Gibbs Smith, 1988.
Dostoevsky, Fyodor. *The Brothers Karamazov*. Translated by Richard Pevear and Larissa Volokhonsky. New York: FSG Books, 2002.
Dunaway, John M. "Estrangement and the Need for Roots: Prophetic Visions of the Human Condition in Albert Camus and Simone Weil." *Religion and Literature* 17, no. 2, *Simone Weil* (1985): 35–42.
Easwaran, Eknath, translator. *Bhagavad Gita*. Boston: Shambhala, 2004.
Elgin, Catherine Z. *Between the Absolute and the Arbitrary*. Ithaca: Cornell University Press, 1997.
Eliot, T.S. "The Love Song of J. Alfred Prufrock." In *Collected Poems, 1909–1935*. London: Faber and Faber, 1936.
Ernstson, Henrik and Erik Swyngedouw. "Framing the Meeting: Rupturing the Anthro-obscene! The Political Promises of Planetary and Uneven Urban Ecologies." At

Conference Teater Reflex, September 16–19, 2015. Organized by KTH Environmental Humanities Laboratory, Stockholm. Position Paper Version 2. Online: www.researchgate.net/profile/Henrik_Ernstson/publication/280738621_Framing_the_Meeting_Rupturing_the_Anthro-obscene_The_Political_Promises_of_Planetary_Uneven_Urban_Ecologies/links/55c4ab2608aebc967df374c4.pdf (accessed October 21, 2016).

Esposito, Roberto. *The Origin of the Political: Hannah Arendt or Simone Weil?* Translated by Vincenzo Binetti and Gareth Williams. New York: Fordham University Press, 2017.

Evernden, Lorne Leslie Neil. *The Natural Alien: Humankind and Environment.* Toronto: University of Toronto Press, 1993.

Farrier, David. *Anthropocene Poetics: Deep Time, Sacrifice Zones, and Extinction.* Minneapolis: University of Minnesota Press, 2019.

Ferrari, Giovanni R.F. *Listening to the Cicadas: A Study of Plato's* Phaedrus. Cambridge: Cambridge University Press, 1987.

Fox, Warwick. "Arne Naess: A Biographical Sketch." *The Trumpeter* 9, no. 2 (1992): 45–49.

———. "On the Interpretation of Naess' Central Term 'Self-realization.'" *The Trumpeter* 7, no. 2 (1990): 98–101.

———. *Toward a Transpersonal Ecology: Developing New Foundations for Environmentalism.* Boston: Shambhala, 1990.

Fox Keller, Evelyn. *A Feeling for the Organism: The Life and Work of Barbara McClintock.* San Francisco: W.H. Freeman, 1983.

Frankl, Viktor E. *Man's Search for Meaning.* Boston: Beacon Press, 2006.

Gabellieri, Emmanuel. "Reconstructing Platonism: The Trinitarian Metaxology of Simone Weil." In *The Christian Platonism of Simone Weil*, 133–158. Edited by E. Jane Doering and Eric O. Springsted. Notre Dame: Notre Dame Press, 2004.

Goethe, Johann Wolfgang von. *Faust: A Tragedy in Two Parts.* Translated by Thomas Wayne. New York: Algora, 2016.

Goodstein, Laurie. "Serenity Prayer Faces Challenge on Authorship." *New York Times* (July 11, 2008). Online: www.nytimes.com/2008/07/11/world/americas/11iht-prayer.4.14439421.html.

Graeber, David and David Wengrow. *The Dawn of Everything: A New History of Humanity.* New York: Picador, 2023.

Gregory, Judith. "A Letter to Simone Weil." *Cross Currents* 40, no. 3 (1990): 368–385.

Grote, Jim. "Prestige: Simone Weil's Theory of Social Force." *Spirituality Today* 42, no. 3 (1990): 217–232.

Hadas, Moses. *A History of Greek Literature.* New York: Columbia University Press, 1953.

Hadot, Pierre. *The Inner Citadel: The Meditations of Marcus Aurelius.* Cambridge, MA: Harvard University Press, 1998.

Hamblin, Madeline. "Simone Weil's Theology of Evil, Love and the Self-Emptying God." In *Mysticism, Nihilism, Feminism: New Critical Essays on the Theology of Simone Weil*, 49–54. Edited by Thomas A. Idinopulos and Josephine Zadovsky Knopp. Johnson City: Institute of Social Sciences and Arts, 1984.

Haraway, Donna J. *Primate Visions: Gender, Race, and Nature in the World of Modern Science.* New York: Routledge, 1989.

———. *Staying with the Trouble.* Durham: Duke University Press, 2016.

Harding, Sandra. "Rethinking Standpoint Epistemology: What Is 'Strong Objectivity?'" In *Feminist Theory: A Philosophical Anthology*, 49–82. Edited by Ann E. Cudd and Robin O. Andreasen. Oxford: Blackwell, 2005.

———. *Whose Science? Whose Knowledge? Thinking From Women's Lives.* Ithaca: Cornell University Press, 1991.

Harrison, Robert Pogue. *Gardens: An Essay on the Human Condition.* Chicago: University of Chicago Press, 2008.

Hedges, Chris. *War Is a Force That Gives Us Meaning.* New York: Public Affairs, 2014.

Hegel, Georg Wilhelm Friedrich. *Phenomenology of Spirit.* Translated by A.V. Miller. Oxford: Oxford University Press, 1977.

Heidegger, Martin. *Basic Writings.* Edited by David Farrell Krell. New York: HarperCollins, 2008.

———. *Being and Time.* Translated by Joan Stambaugh. Edited by Dennis J. Schmidt. Albany: SUNY Press, 2010.

Heldke, Lisa M. and Stephen H. Kellert. "Objectivity as Responsibility." *Metaphilosophy* 26, no. 4 (1995): 360–378.

Herbert, George. *George Herbert and the Seventeenth-Century Religious Poets.* Edited by Mario Di Cesare. New York: W.W. Norton, 1978.

Hermida, J. Ranilo B. "Simone Weil: A Sense of God." *Logos: A Journal of Catholic Thought and Culture* 9, no. 1 (2006): 127–144.

Hermsen, Joke J. "The Impersonal and the Other: On Simone Weil." *European Journal of Women's Studies* 6, no. 2 (1999): 183–200.

Hinshelwood, Brad. "The Carolinian Context of John Locke's Theory of Slavery." *Political Theory* 41, no. 4 (2013): 562–590.

Hird, Myra J. *The Origins of Sociable Life: Evolution after Science Studies.* Basingstoke: Palgrave Macmillan, 2009.

Hobbes, Thomas. *Leviathan.* Baltimore: Penguin Books, 1968.

Holifield, Ryan, Jayajit Chakraborty, and Gordon Walker, editors. *The Routledge Handbook of Environmental Justice.* New York: Routledge, 2020.

Homer. *The Odyssey.* Translated by Robert Fagles. In *The Norton Anthology of Western Literature*, 8th edition, vol. 1, 206–495. Edited by Peter Simon. New York: W.W. Norton, 2006.

Honig, Bonnie. *Political Theory and the Displacement of Politics.* Ithaca: Cornell University Press, 1993.

hooks, bell. *Belonging: A Culture of Place.* Abingdon: Routledge, 2008.

———. *Outlaw Culture: Resisting Representations.* Abingdon: Routledge, 2015.

Huesemann, Michael and Joyce Huesemann, *Techno-Fix: Why Technology Won't Save Us or the Environment.* Guelph: New Society Publishers, 2011.

Hume, David. *A Treatise of Human Nature: An Attempt to Introduce the Experimental Method of Reasoning into Moral Subjects.* Oxford: Oxford University Press, 1978.

Irigaray, Luce and Michael Marder. *Through Vegetal Being: Two Philosophical Perspectives.* New York: Columbia University Press, 2016.

Jesson, Stuart. "'The Question in Each and Every Thing': Nietzsche and Weil on Affirmation." *International Journal for Philosophy of Religion* 86, no. 2 (2019): 131–155.

Kant, Immanuel. *Critique of Pure Reason.* Translated and edited by Paul Guyer and Allen W. Wood. Cambridge: Cambridge University Press, 1999.

Keller, Catherine. *Political Theology of the Earth: Our Planetary Emergency and the Struggle for a New Public.* New York: Columbia University Press, 2018.

———. *The Face of the Deep: A Theology of Becoming.* New York: Routledge, 2003.

Keller, Catherine and Mary-Jane Rubenstein, editors. *Entangled Worlds: Religion, Science, and New Materialisms.* New York: Fordham University Press, 2017.

Kimmerer, Robin Wall. *Braiding Sweetgrass: Indigenous Wisdom, Scientific Knowledge, and the Teachings of Plants.* Minneapolis: Milkweed Editions, 2013.

Klein, Naomi. *The Shock Doctrine.* Toronto: Vintage Books, 2007.

———. *This Changes Everything: Capitalism vs. the Climate*. New York: Knopf Canada, 2014.
Kostigen, Thomas M. *Hacking Planet Earth: How Geoengineering Can Help Us Reimagine the Future*. New York: Tarcher, 2020.
Kotva, Simone. *Effort and Grace: On the Spiritual Exercise of Philosophy*. London: Bloomsbury, 2020.
Leopold, Aldo. "Land Ethic." In *A Sand County Almanac: With Other Essays on Conservation from Round River*, 217–240. Edited by Charles Walsh Schwartz. New York: Oxford University Press, 1966.
Levinas, Emmanuel. *Otherwise Than Being or Beyond Essence*. Translated by Alphonso Lingis. Pittsburgh: Duquesne University Press, 1998.
———. "Simone Weil Against the Bible." In *Difficult Freedom: Essays on Judaism*, 133–141. Translated by Sean Hand. London: Athlone Press, 1990.
———. *Totality and Infinity: An Essay on Exteriority*. Translated by Alphonso Lingis. Pittsburgh: Duquesne, 2013.
Lewis, C.S. *A Grief Observed*. New York: HarperOne, 2015.
Linck, Matthew S. "Unmastering Speech: Irony in Plato's 'Phaedrus.'" *Philosophy and Rhetoric* 36, no. 3 (2003): 264–276.
Lloyd, Genevieve. "The Divided Soul: Manliness and Effeminacy." In *The Man of Reason: "Male" and "Female" in Western Philosophy*, 18–38. Minneapolis: University of Minnesota Press, 1993.
Locke, John. *Second Treatise of Government*. Edited by Dave Gowan and Chuck Greif. Project Gutenberg, 2003. Online: EBook #7370.
Luik, John C. "Humanism." In *Routledge Encyclopedia of Philosophy*. Abingdon: Taylor and Francis, 1998. https://doi.org/10.4324/9780415249126-N025-1.
Luke, Timothy W. *Ecocritique: Contesting the Politics of Nature, Economy, and Culture*. Minneapolis: University of Minnesota Press, 1997.
MacFarlane, Robert. *Landmarks*. London: Hamish Hamilton, 2015.
Matthews, Freya. *The Ecological Self*. Abingdon: Taylor and Francis, 1991.
McCullough, Lissa. "Simone Weil's Phenomenology of the Body." *Comparative and Continental Philosophy* 4, no. 2 (2012): 195–218.
———. "The Political Import of Weil's Religious Turn." At Hannah Arendt and Simone Weil Conference, April 16–18, 2021.
———. *The Religious Philosophy of Simone Weil: An Introduction*. New York: I.B. Tauris, 2014.
McFague, Sallie. *The Body of God: An Ecological Theology*. Minneapolis: Fortress, 1993.
McKibben, Bill. "The End of Nature." In *The Norton Book of Nature Writing*, 1120–1130. Edited by Robert Finch and John Elder. New York: W.W. Norton, 2002.
———. *The End of Nature*. New York: Random House, 1989.
McLellan, David. *Utopian Pessimist*. Berkeley: Poseidon Press, 1990.
Meaney, Marie Cabaud. *Simone Weil's Apologetic Use of Literature: Her Christological Interpretations of Ancient Greek Texts*. Oxford: Oxford University Press, 2007.
Mele, Alfred R. "Self-Control, Action, and Belief." *American Philosophical Quarterly* 22, no. 2 (1985): 169–175.
Meltzer, Françoise. "The Hands of Simone Weil." *Critical Inquiry* 27, no. 4 (2001): 611–628.
Merchant, Carolyn. *The Death of Nature: Women, Ecology, and the Scientific Revolution*. New York: HarperOne, 1990.
Mills, Charles W. *The Racial Contract*. Ithaca: Cornell University Press, 2022.
Miłosz, Czesław. "The Importance of Simone Weil." In *On the Abolition of All Political Parties*. Translated by Simon Leys. New York: New York Review of Books, 2013.

Monahan, Michael J. "Emancipatory Affect: bell hooks on Love and Liberation." *CLR James Journal: Special Issue on the Emancipatory Thought of bell hooks* 17, no. 1 (2011): 102–111.
Montaigne, Michel. *Essays of Michel de Montaigne.* Translated by Charles Cotton. Edited by William Carew Hazlitt. Project Gutenberg, 2006. Online: EBook #3600.
Morgan, Vance G. *Weaving the World: Simone Weil on Science, Mathematics, and Love.* Notre Dame: Notre Dame University Press, 2005.
Moore, Jason W. *Anthropocene or Capitalocene? Nature, History, and the Crisis of Capitalism.* Oakland: PM Press, 2016.
Morton, Oliver. *The Planet Remade: How Geoengineering Could Change the World.* Princeton: Princeton University Press, 2017.
Mouffe, Chantal. *Agonistics: Thinking the World Politically.* New York: Verso, 2013.
Murdoch, Iris. *The Sovereignty of Good.* London: Routledge, 2001.
Mutual Aid Katarokwi/Kingston. Kingston, ON: AKA Autonomous Social Center. Online: https://mutualaidkatarokwi.wordpress.com/.
Naess, Arne. "Self-Realization: An Ecological Approach to Being in the World." *The Trumpeter: Voices of the Canadian Eco-philosophy Network* 4, no. 3 (1987): 35–42.
———. "Self-Realization in Mixed Communities of Humans, Bears, Sheep, and Wolves." *Inquiry* 22 (1979): 231–241.
Nash, Jennifer C. "Practicing Love: Black Feminism, Love-Politics, and Post-Intersectionality." *Meridians* 11, no. 2 (2011): 1–24.
Nell, Edward, editor. *Free Market Conservatism (Routledge Revivals): A Critique of Theory and Practice*, 1st edition. Florence: Taylor and Francis, 2009.
Nietzsche, Friedrich. *The Gay Science.* Translated by Thomas Common. Mineola: Dover Publications, 2006.
———. *Thus Spoke Zarathustra.* Translated by R.J. Hollingdale. New York: Penguin Books, 2003.
Okin, Susan Moller. *Women in Western Political Thought.* Princeton: Princeton University Press, 2013.
Olsson, Karen. *The Weil Conjectures: On Math and the Pursuit of the Unknown.* New York: Farrar, Straus and Giroux, 2019.
"Overrepresentation of Indigenous People in the Canadian Criminal Justice System: Causes and Responses." *Reports and Publications*. Research and Statistics Division, Department of Justice, Government of Canada, 2020. Online: www.justice.gc.ca/eng/rp-pr/jr/oip-cjs/e.html.
Oxford English Dictionary Online. March 2022. Oxford University Press.
———. "irony, n." www-oed-com.proxy.queensu.ca/view/Entry/99565?rskey=x98c1yandresult=1 (accessed April 6, 2022).
———. "nature, n.," 11. www-oed-com.proxy.queensu.ca/view/Entry/125353?result=1andrskey=gPwZ4Vand (accessed April 13, 2021).
———. "nature, n.," 5a, b. www-oed-com.proxy.queensu.ca/view/Entry/125353?result=1andrskey=gPwZ4Vand (accessed April 13, 2021).
———. "Socratic, adj. and n." www-oed-com.proxy.queensu.ca/view/Entry/183829?redirectedFrom=socratic+irony (accessed April 6, 2022).
Palsson, Gisli and Heather A. Swanson. "Down to Earth: Geosocialities and Geopolitics." *Environmental Humanities* 8, no. 2 (2016): 149–171.
Pateman, Carole. *The Sexual Contract.* Stanford: Stanford University Press, 2018.
Peck, M. Scott. *The Different Drum: Community Making and Peace.* New York: Touchstone, 1998.

Pétrement, Simone. *Simone Weil: A Life*. Translated by Raymond Rosenthal. New York: Pantheon Books, 1976.
Pickstock, Catherine. *After Writing: On the Liturgical Consummation of Philosophy*. Oxford: Blackwell, 1998.
Pirruccello, Ann. "Making the World My Body: Simone Weil and Somatic Practice." *Philosophy East and West* 52, no. 4 (2002): 479–497.
Plant, Stephen. *Simone Weil: A Brief Introduction*. London: HarperCollins, 2007.
Plato. *Phaedo*. Translated by David Gallop. Oxford: Clarendon Press, 1975.
———. *Phaedrus*. Translated by Benjamin Jowett. Bulgaria: Demetra, 1870.
———. *Republic*. Translated by G.M.A. Grube. Indianapolis: Hackett, 1874.
———. *Republic*. Translated by James Adam. Cambridge: Cambridge University Press, 1963.
———. *Republic*. Translated by Francis Cornford. Oxford: Oxford University Press, 1965.
———. *Symposium*. Edited by M.C. Howatson and Frisbee C.C. Sheffield. Translated by M.C. Howatson. Cambridge: Cambridge University Press, 2008.
Plessix Gray, Francine du. "At Large and at Small: Loving and Hating Simone Weil." *American Scholar* 70, no. 3 (2001): 5–11.
Plumwood, Val. *Feminism and the Mastery of Nature*. New York: Routledge, 1993.
Pomeroy, Sarah. "Feminism in Book V of Plato's *Republic*." *Apeiron* 8, no. 1 (1974): 32–35.
Pseudo-Dionysius. "The Divine Names." In *Pseudo-Dionysius: The Complete Works*, 47–132. Translated by Colm Luibheid. New York: Paulist Press, 1987.
Purcell, Donald. "Iris Murdoch's *The Green Knight* and Simone Weil." *Cahiers Simone Weil* 19, no. 2 (1996): 225–238.
Rabinowitz, Ann. "Tzimtzum." *Judaism: A Quarterly Journal of Jewish Life and Thought* 47, no. 2 (1998): 213.
Ramshaw, Gail. *God Beyond Gender: Feminist Christian God-Language*. Minneapolis: Augsburg Press, 1995.
Rapp, Jennifer R. *Ordinary Oblivion and the Self Unmoored: Reading Plato's* Phaedrus *and Writing the Soul*. New York: Fordham University Press, 2014.
Reed, Robert Charles. "Decreation as Substitution: Reading Simon Weil Through Levinas." *Journal of Religion* 93, no. 1 (2013): 25–40.
Reps, Paul and Nyogen Senzaki, editors. *Zen Flesh, Zen Bones: A Collection of Zen and Pre-Zen Writings*. Rutland: Tuttle, 1985.
Rihll, T.E. "Teaching and Learning in Classical Athens." *Greece and Rome* 50, no. 2 (2003): 168–190.
Rilke, Rainer Maria. *The Selected Poetry of Rainer Maria Rilke*. Translated by S. Mitchell. New York: Random House, 1982.
Ritner, Scott B. "A Critique of Greatness." *Theory & Event* 26, no. 2 (2023): 345–367.
Rozelle-Stone, A. Rebecca and Lucian Stone. *Simone Weil and Theology*. London: Bloomsbury, 2013.
Salinger, J.D. *The Catcher in the Rye*. New York: Little Brown, 1951.
Sallis, John. *The Return of Nature: On the Beyond of Sense*. Indianapolis: Indiana University Press, 2016.
Schmidt, Lawrence. "The Language of Limitation as the Key to Simone Weil's Understanding of Beauty and Justice." *Attention: The Life and Legacy of Simone Weil* 1, no. 2 (2021). https://attentionsw.org/the-language-of-limitation-as-the-key-to-simone-weils-understanding-of-beauty-and-justice/.
Scott, Gary Alan and William A. Welton. *Erotic Wisdom: Philosophy and Intermediacy in Plato's Symposium*. Albany: State University of New York Press, 2008.

Seed, John. "The Ecological Self." *The Trumpeter* 22, no. 2 (2006): 96–102.
Shapin, Steven. *A Social History of Truth: Civility and Science in Seventeenth-Century England*. Chicago: University of Chicago Press, 1994.
———. *Never Pure: Historical Studies of Science As If It Was Produced by People with Bodies, Situated in Time, Space, Culture, and Society, and Struggling for Credibility and Authority*. Baltimore: Johns Hopkins University Press, 2010.
Shiffman, Mark G. Review of *Weaving the World: Simone Weil on Science, Mathematics, and Love*, by Vance G. Morgan. *Notre Dame Philosophical Reviews* (2006). Online: https://ndpr.nd.edu/reviews/weaving-the-world-simone-weil-on-science-mathematics-and-love/.
Shotwell, Alexis. *Against Purity: Living Ethically in Compromised Times*. Minneapolis: University of Minnesota Press, 2016.
Smith, Mick. *Against Ecological Sovereignty: Ethics, Biopolitics, and Saving the Natural World*. Minneapolis: University of Minnesota Press, 2011.
———. *An Ethics of Place: Radical Ecology, Postmodernity, and Social Theory*. New York: SUNY Press, 2001.
———. "Deep Ecology: What Is Said and (to Be) Done?" *The Trumpeter* 30, no. 2 (2014): 141–156.
———. "Dis(appearance): Earth, Ethics and Apparently (In)Significant Others." *Australian Humanities Review*, 50 (2011): 23–44.
———. "Environmental Anamnesis: Walter Benjamin and the Ethics of Extinction." *Environmental Ethics* 23, no. 4 (2001): 359–376.
Sophocles. Antigone. Translated by Elizabeth Wyckoff. In *Greek Tragedies I*, 187–240. Edited by David Grene and Richard Lattimore (Third edition editors Mark Griffith and Glenn W. Most). Chicago: University of Chicago Press, 2013.
Spinoza, Benedict de. *The Ethics*. Translated by R.H.M. Elwes. Project Gutenberg, 2003. Online: EBook #3800.
Springsted, Eric O. *Christus Mediator: Platonic Mediation in the Thought of Simone Weil*. Chico: Scholars Press, 1983.
———. "Of Tennis, Persons and Politics." *Philosophical Investigations* 16, no. 3 (1993): 198–211.
Stadler, Friedrich. *The Vienna Circle. Studies in the Origins, Development and Influence of Logical Empiricism*. Translated by Camilla Nielson, Joel Golb, Sabine Schmidt, and Thomas Ernst. Vienna and New York: Springer, 1997.
Steiner, George. "Sainte Simone – Simone Weil." In *No Passion Spent*, 171–179. New Haven: Yale University Press, 1996.
Sullivan, Heather I., and Bernhard F. Malkmus. "The Challenge of Ecology to the Humanities: An Introduction." *New German Critique* 43, no. 2 (2016): 1–20.
Tarbuck, Alice and Simone Kotva. "The Non-Secular Pilgrimage: Walking and Looking in Ken Cockburn and Alec Finlay's *The Road North*." *Critical Survey* 29, no. 1 (2017): 33–52.
Taubes, Susan Anima. "The Absent God." *Journal of Religion* 35, no. 1 (1955): 6–16.
Trelawny-Cassity, Lewis. "Plato: The Academy." In *The Internet Encyclopedia of Philosophy*. ISSN 2161-0002. Online: https://iep.utm.edu/.
Truth and Reconciliation Commission of Canada. National Centre for Truth and Reconciliation, University of Manitoba, 2015. Online: https://nctr.ca/records/reports/#trc-reports.
Tsing, Anna Lowenhaupt. "On Nonscalability: The Living World Is Not Amenable to Precision-Nested Scales." *Common Knowledge* 18, no. 3 (2012): 505–524.

———. *The Mushroom at the End of the World: On the Possibility of Life in Capitalist Ruins*. Princeton: Princeton University Press, 2015.

Tuana, Nancy. *Feminist Interpretations of Plato*. University Park: Pennsylvania State University Press, 1994.

Veto, Miklos J. "Simone Weil and Suffering." *Thought: Fordham University Quarterly* 40, no. 2 (1965): 275–286.

Vlastos, Gregory. "Socrates Contra Socrates in Plato." In *Socrates: Ironist and Moral Philosopher*, 45–80. Cambridge: Cambridge University Press, 1991.

Vries, Hent de. "Introduction." In *Political Theologies: Public Religions in a Post-Secular World*, 1–88. Edited by Hent de Vries and Lawrence Sullivan. New York: Fordham University Press, 2006.

Waller, Donald M. "Getting Back to the Right Nature: A Reply to Cronon's 'The Trouble with Wilderness.'" In *The Great New Wilderness Debate*, 540–567. Edited by J.B. Callicott and M.P. Nelson. Atlanta: University of Georgia Press, 1998.

Warren, Karen J. "Second Kind of Position in Feminist Environmental Philosophy: Ecofeminist Philosophy." In "Feminist Environmental Philosophy," *The Stanford Encyclopedia of Philosophy* (2015). Edited by Edward N. Zalta. Online: https://plato.stanford.edu/entries/feminism-environmental/#SecKinPosFemEnvEco.

Watson, Gary. "Skepticism About Weakness of Will." *Philosophical Review* 86, no. 3 (1977): 316–339.

Weil, Simone. *First and Last Notebooks*. Translated by Richard Rees. Eugene: Wipf and Stock, 1970.

———. *Gravity and Grace*. Edited by Gustave Thibon and Thomas R. Nevin. Translated by Arthur Wills. Lincoln: University of Nebraska Press, 1997.

———. "Human Personality." In *Simone Weil: An Anthology*, 49–78. London: Penguin Books, 2005.

———. *Intimations of Christianity Among the Ancient Greeks*. London: Ark Paper Backs, 1957.

———. *Lectures on Philosophy*. Translated by Hugh Price. Cambridge: Cambridge University Press, 1978.

———. *Love in the Void: Where God Finds Us*. Edited by Laurie Gagne. Walden: Plough, 2018.

———. *On the Abolition of all Political Parties*. Translated by Simon Leys. New York: New York Review of Books, 2013.

———. "Science et Perception dans Descartes." In *Sur la science*, 11–99. Paris: Gallimard, 1966.

———. *Selected Essays, 1934–1943: Historical, Political, and Moral Writings*. Edited and translated by Richard Rees. Eugene: Wipf and Stock, 1962.

———. *Seventy Letters*. Translated by Richard Rees. Eugene: Wipf and Stock, 1969.

———. "Some Reflections Around the Concept of Value: On Valéry's Claim that Philosophy Is Poetry." Translated by Eric O. Springsted. *Philosophical Investigations* 37, no. 2 (2014): 105–112.

———. "The Iliad, or the Poem of Force." Translated by Mary McCarthy. In *Simone Weil: An Anthology*, 182–215. London: Penguin Books, 2005.

———. *The Need for Roots*. Translated by Arthur Wills. New York: Routledge, 1952.

———. *The Notebooks of Simone Weil*. Translated by Arthur Wills. London: Routledge, 2004.

———. *Waiting for God*. Translated by Emma Craufurd. New York: HarperCollins, 2009.

Whitman, Walt. "Song of Myself." In *Leaves of Grass*. New York: Norton, 1973. Poetry Foundation. Online: www.poetryfoundation.org/poems/45477/song-of-myself-1892-version.

Williams, Rowan. "Review of *The Just Balance* by Peter Winch." *Philosophical Investigations* 14, no. 2 (1991): 155–171.

———. "The Necessary Nonexistence of God." In *Simone Weil's Philosophy of Culture: Readings Toward a Divine Humanity*, 52–77. Edited by R.H. Bell. Cambridge: Cambridge University Press, 1993.

Winch, Peter. *Simone Weil: The "Just Balance."* Cambridge: Cambridge University Press, 1989.

Wolfteich, Claire. "Attention or Destruction: Simone Weil and the Paradox of the Eucharist." *Journal of Religion* 81, no. 3 (2001): 359–376.

Zimmerman, Michael. *Heidegger's Confrontation with Modernity: Technology, Politics, and Art*. Bloomington: Indiana University Press, 1990.

Zwicky, Jan. "Lyric Realism: Nature, Poetry, Silence, and Ontology." Green Imagination Issue. *Malahat Review* 165 (Winter 2008): 85–91.

Index

Note: Endnotes are indicated by the page number followed by "n" and the note number e.g., 117n11 refers to note 11 on page 117. Page numbers in *italic* refer to Figures.